W9-AQO-765

# *The* NATURAL HISTORY *of the* OAK TREE

# *The* NATURAL HISTORY *of the* OAK TREE

RICHARD LEWINGTON
& DAVID STREETER

DORLING KINDERSLEY
LONDON • NEW YORK • STUTTGART

*Dedicated to Jack and Beryl Lewington, and to James, Katharine, and Olivia Streeter*

A DORLING KINDERSLEY BOOK

Managing Editor: Krystyna Mayer
Managing Art Editor: Derek Coombes
Production Controller: Antony Heller

**CREDITS AND ACKNOWLEDGMENTS**

Illustrations on pages 20 to 25 by Ian Lewington.

Dorling Kindersley would like to thank the following people for their help in the preparation of this book: Candida Ross-Macdonald, Lynn Parr, and Sharon Lucas for editorial assistance; Colette Ho for design assistance; Patrizio Semproni and Doug Miller for computer page make-up.

Richard Lewington would like to thank the following for their help: Michael J. Amphlett; George Bloom, Abingdon Naturalists Society; Patricia Jane Dart; Michael Chinery; Christopher Leach, The British Plant Gall Society; Georgina Lewington; George McGavin, Oxford University Museum; Tony Rayner.

David Streeter wishes to thank Jenny Long for her care in typing the text.

First American Edition 1993
2 4 6 8 10 9 7 5 3 1

Published in the United States by
Dorling Kindersley, Inc., 232 Madison Avenue
New York, New York 10016

**Library of Congress Cataloging-in-Publication Data**

Lewington, Richard.
The natural history of the oak tree : an intricate visual exploration of the oak and its environment / by Richard Lewington, David Streeter.
— 1st American ed.
p. cm.
ISBN 1-56458-307-4
1. Forest ecology—Great Britain—Pictorial works. 2. English oak—Great Britain—Ecology—Pictorial works. 3. Natural History—Great Britain—Pictorial works. I. Streeter, David, B. Sc. II. Title
QH137.L615 1993
574.5'2642'0941—dc20 93-18573
CIP

Reproduced in Singapore by Koford pte
Printed and bound in Italy by New Interlitho SPA

# CONTENTS



# PREFACE

Oak trees are found throughout the temperate regions of the Northern Hemisphere, from southern Scandinavia to Japan, and from North America's eastern states to the Pacific coast. In all, about 450 species have been described, about half of which are evergreen. Twenty-five occur in Europe, two of which – the Common Oak and the Sessile Oak – extend from southern Norway to Sicily and from Ireland to the Urals.

This book describes the natural history of these two oaks, the only species to occur naturally in northern Europe. The book is about trees rather than woods, and describes the large and complex community of fungi, plants, and animals that depend on the oak for their existence. More different kinds of insect are found on the oak than on any other European plant. In spring the canopy is alive with foraging birds, and a single tree may have more than 30 different species of lichen growing on its bark. In addition, a vast array of fungi and invertebrates live on its dead and decaying wood and in the layer of leaf litter shed by the trees each autumn.

The colour plates took more than two years to prepare, and almost all the specimens were painted from life. Many were found in Richard Lewington's local Oxfordshire oakwood, Fulscot Wood. Others had to be sought further afield in some of Britain's most famous woodlands, such as Oxford University's Wytham Wood, the ancient wood pastures of the New Forest in Hampshire, Windsor Great Park, and Wistman's Wood in the Dartmoor National Park. Many of the gall wasps and micro moths are extremely difficult to collect, and had to be painstakingly bred from their galls and leaf mines.

Throughout history, the oak has played a special role in the social, economic, and religious lives of the peoples of north-west Europe. We hope that this book will help lift the veil on the day-to-day life of the oak tree, and strengthen our resolve to cherish our woodlands.

# The Common Oak

MUCH OF EUROPE WAS ONCE covered by woodland. Selective exploitation of the "wildwood" in antiquity resulted in the oak becoming the most common and characteristic tree over most of north-west Europe, including Britain, where it has come to symbolize the English countryside. Throughout Europe the oak was a sacred tree, associated with the gods Zeus and Thor.

The oaks of northern and central Europe belong to two closely related species: the Common or Pedunculate Oak, which has stalked acorns, and the Sessile Oak, which has acorns without stalks (*see pages 8-9*). The Common Oak, *Quercus robur*, is found from Norway to the Mediterranean, and east to the Urals. It grows best on heavy, fertile soils of lowland clays and the alluvium of river valleys, although it also grows on the bleak granite uplands of the Dartmoor National Park, in southwest Britain. Oak is the most durable of timbers, and was used exclusively in medieval timber-framed buildings and for shipbuilding. The Common Oak was the most favoured, because its crooked branches provided the curved timbers needed for the supporting trusses and braces.

That oaks live to a great age is well known. In Europe, specimens up to 450 years old are well authenticated, while one tree in Switzerland yielded a ring count of 930 years. In Britain most old oaks are not more than 250 years old.

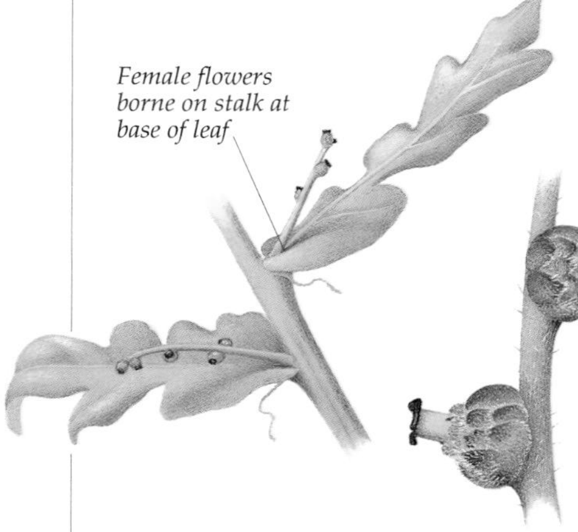

*Female flowers borne on stalk at base of leaf*

*Female flower with receptive stigmas*

**FEMALE FLOWERS IN SPRING**
The female flowers of the Common Oak are borne on stalked spikes that arise from the axils of leaves, the points at which the leaves join the twigs. Each spike bears up to five flowers, and each flower is surrounded at its base by a ring of overlapping scales that later develops into the cup of the acorn.

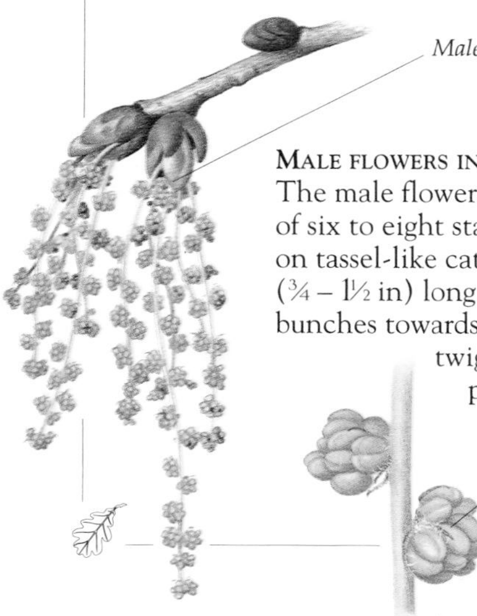

*Male catkins*

**MALE FLOWERS IN SPRING**
The male flowers, which consist of six to eight stamens, are borne on tassel-like catkins about 2–4 cm (¾ – 1½ in) long. These arise in bunches towards the ends of young twigs. When ripe, pollen is released on the wind.

*Male flower*

**FAMILIAR SHAPE**
The trunk of the Common Oak tends to disappear into the crown, and the boughs are irregularly branched, bent, and crooked. Woodland trees grow taller than those in the open. In Europe, trees up to 45 metres (150 ft) in height are not uncommon, although those in Great Britain do not usually grow over 27 metres (90 ft) tall.

**LEAVES IN SUMMER**
The leaves of the Common Oak are rather dull green in colour above, and have three to five pairs of lobes. The stalk is short, usually less than 5 mm (¼ in) long. The base of the leaf blade forms two small lobes, or auricles. The undersurface of the leaf is paler, and completely without hairs.

**ACORNS IN AUTUMN**
The acorn is the fruit of the oak and contains the seed. It is surrounded at its base by a hard, warty cup, or cupule. The acorns of the Common Oak are borne on stalks up to 8 cm (3 in) long. Oaks first produce acorns when they are 40 to 50 years old.

**TWIG IN WINTER**
Although the tree is dormant in winter, with no new growth, the young, undeveloped leaves for the next spring lie protected from cold and wind within an envelope of rusty-brown bud scales. The leaf buds of the oak are characteristically clustered at the tips of the branches.

# The Sessile Oak

The Sessile Oak, or Durmast Oak, *Quercus petraea*, has a similar distribution in Europe to the Common Oak, but does not extend as far east into Russia. In Britain, it is most common on hills in the west. It prefers well-drained, acidic soils to the heavy clays and alluviums favoured by the Common Oak, but in some woods the two species grow together.

In parts of northern Europe, Sessile Oak coppice comprises much of the oak woodland. Coppicing is a traditional form of woodland management, in which the trees are cut to the ground or near ground level at regular intervals. This stimulates the plant to produce numerous young shoots, or poles, from the stump, or stool. In the past, the frequency of oak coppicing depended upon the desired use of the poles. Tannin could be extracted by soaking the bark, and as the amount of tannin in bark decreases with age, the coppice rotation for tannin production was kept to about 15 years. A 20-year rotation was preferred for charcoal production, and if the wood was to be used for pit props or fuel, the rotation would be extended to 30 years. Much of the old Sessile Oak coppice that remains on hillsides in the west of Britain is no longer managed in the traditional way.

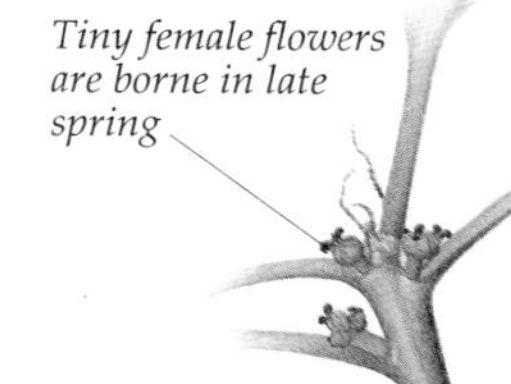


*Stigmas for accepting pollen from male flowers*

**Female flowers in spring**
The female flowers are almost stalkless, and are borne in small groups on twigs, in the axils of leaves. Pollen adheres to the three receptive stigmas on the tip of each flower.

**Male flowers in spring**
The male catkins shed their pollen during late spring, and each catkin produces several million pollen grains. Oak trees are probably self-sterile (the pollen cannot fertilize female flowers from the same tree). Several species of gall wasp cause galls on the catkins, of which the most familiar is the "currant gall" of *Neuroterus quercusbaccarum* *(see page 43)*.

*Male flowers*

**Leaves in summer**
Between five and eight lobes lie on each side of the glossy, green leaves. The lower surface of each leaf is conspicuously hairy, especially along the midrib and in the axils of the lower veins. The base of the leaf generally tapers into the leaf stalk, which is usually between 10 mm (½ in) and 20 mm (1 in) long.

**Acorns in autumn**
The acorns of the Sessile Oak are either stalkless, or have very short stalks no more than 3 mm (⅛ in) long. The Sessile Oak fruits less frequently than the Common Oak, and its acorns are slightly smaller.

**Twig in winter**
The buds of the Sessile Oak are larger and more pointed than those of the Common Oak, and the rings of leaf scars at a shoot's base show the extent of spring and summer growth. Some moth larvae, such as those of the Light Emerald, *Campaea margaritata*, feed on the bark of young twigs throughout the winter.

**The canopy shape**
The branches of the Sessile Oak are straighter than those of the Common Oak, and the main trunk is less branched.

# From Acorn to Tree

A single oak tree produces several million acorns in its life. Very few of these survive to produce acorns themselves. While still on the tree, acorns are attacked by weevils and moth caterpillars, and consumed in enormous quantities by squirrels and Wood Pigeons. On the ground, they are eaten by rodents and deer. In mainland Europe, acorns are also an important part of the diet of wild boars. In winter, acorns that have escaped these dangers may be infected by the fungus *Ciboria batschiana*, which turns the seed into a hard, black mass. Those that survive to germinate need the right conditions. Light shade is best for oaks – but too much shade makes them susceptible to mildew *(see page 50)*. The young seedlings are eagerly sought by browsing animals, and those under a tree canopy suffer the spring rain of hungry caterpillars *(see page 14)*.

Mid-summer (shown x 1½)

Late summer (shown x 1½)

Early autumn (shown x 1½)

Mid-autumn (shown x 1½)

**Acorn growth**
In late spring, the oak's female flowers are pollinated by the wind with pollen from the male flowers, and acorns begin to form. Initially, each acorn may contain up to six seeds, but usually only one of these develops. The ring of overlapping scales around the base of the female flower grows with the acorn, and eventually hardens to form the familiar cup, or cupule. By autumn, the acorns are fully grown, and beginning to ripen and turn brown.

*Plumule grows straight up towards light*

**Germination**
An oak sheds its acorns before leaf-fall, so the fallen acorns are soon buried under a thick blanket of leaves. This protects them from frost, and hides them from foraging animals. The acorns of the Sessile Oak begin to germinate almost immediately, those of the Common Oak after a short delay. The coat of the acorn is split by the young root, or radicle, which emerges and grows rapidly downwards.

**Early growth**
A few acorns remain dormant until the spring, but by the end of winter most of them have tap roots 10 to 20 cm (4 to 8 in) long. By late spring the young shoot, or plumule, appears above ground. The shoot grows a few small-scale leaves before the first true foliage leaves develop.

*First root grows directly downwards*

**First summer**
If the young shoot survives the annual rain of caterpillars from the canopy above, the plumule continues to elongate. The shoot normally remains unbranched in the first year, but by mid-summer it will have grown five or six foliage leaves and developed a terminal bud at its top. Before the end of the growing season, a second shoot develops from this bud. At this stage, the seedling is particularly vulnerable to browsing animals.

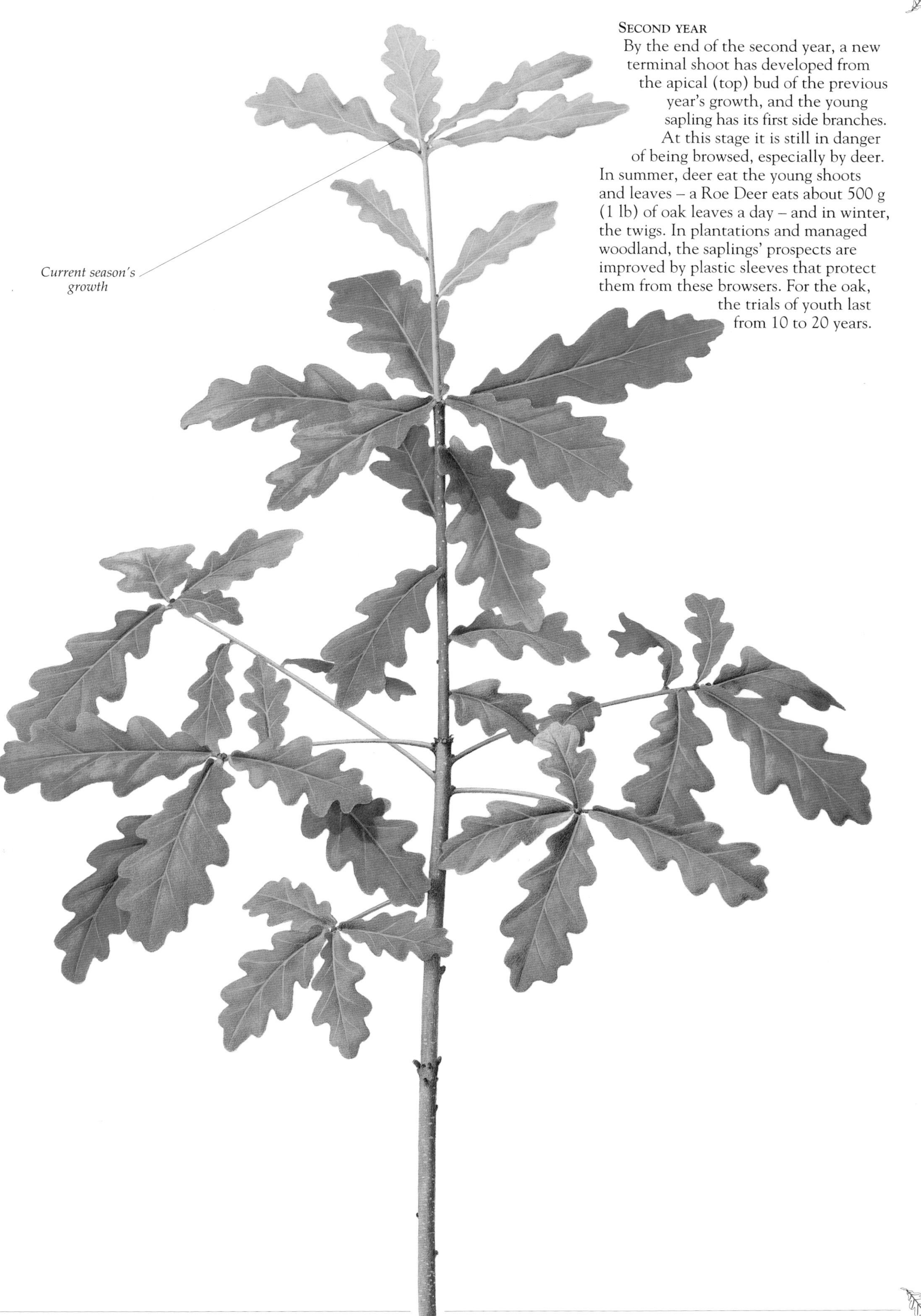


**SECOND YEAR**

By the end of the second year, a new terminal shoot has developed from the apical (top) bud of the previous year's growth, and the young sapling has its first side branches. At this stage it is still in danger of being browsed, especially by deer. In summer, deer eat the young shoots and leaves – a Roe Deer eats about 500 g (1 lb) of oak leaves a day – and in winter, the twigs. In plantations and managed woodland, the saplings' prospects are improved by plastic sleeves that protect them from these browsers. For the oak, the trials of youth last from 10 to 20 years.

# The Oak in Winter

In winter, the oakwood seems still and lifeless, with the skeleton-like fingers of the tree branches standing out starkly against the grey skies. But even during the shortest days there is much activity. Bands of tits flit high in the canopy seaching the dormant buds and crevices in the bark for larvae and other invertebrates. On mild days, the female Robin proclaims her territory with her winter song. Both the Grey and the Red Squirrels are active even during the coldest weather, busily searching for hoards of acorns and hazelnuts randomly buried earlier in the autumn. A fall of fresh snow reveals the characteristic broad, five-toed imprint of the Badger, while the footprints, or slots, of deer show where a family group has crossed a stream. Of the woodland mammals only the bats, the Hedgehog, and the Dormouse undergo true hibernation. Some insects, too, are active during the winter. In early winter, the Winter Moths emerge from their pupal cases in the leaf litter. At dusk, the wingless females crawl up the trunks of trees, where they attract the winged males. After mating, and under the cover of darkness, they continue up the trunk, each laying about 150 eggs in crevices high up in the bark, or hidden among lichens, ready to hatch just as the young leaf buds burst in spring. Mosses and lichens are especially noticeable at this time of year, since they remain vigorous throughout the winter.

**Winter Moth**
*(See page 32)*
The Winter Moth is one of the most common woodland moths. The adults emerge in early winter.

***Parmelia caperata***
*(See page 54)*
A very distinctive and common lichen in unpolluted areas, *Parmelia caperata* forms extensive, yellowish-grey patches on tree trunks and branches.

**The branch in winter**
The leafless branch of the oak can be recognized in winter by its oval, polished, light-brown, hairless buds, clustered together at the tips of the twigs.

MARBLE GALLS
*(See page 40)*
The familiar marble galls, or oak nuts, are caused by the gall wasp *Andricus kollari*. The galls were introduced to Devon from the Middle East in about 1830 for dyeing cloth and making ink.

THE TREE IN WINTER
Growth ceases during the winter because there are no leaves to carry out photosynthesis. The annual cycle of growth is reflected in the annual rings of the trunk, the winter period appearing as a ring separating the summer wood of the previous year from the spring wood of the next. The number of annual rings on a cut tree stump is equivalent to the age of the tree. Since the width of a ring is largely dependent on that year's weather, archaeologists can use the rings to date wooden objects.

# THE OAK IN SPRING

IN SPRING, THE OAK TREE is a site of intense activity. The young leaves break free from the protection of their buds, and the number of caterpillars in the foliage is so great that on a fine day, their droppings, or frass, falling on to the ground can sound like rain.

The oak has evolved a sophisticated chemical defence mechanism that helps to protect it from leaf-eating caterpillars. As the leaf develops, it manufactures large amounts of tannin, a chemical that not only has a repellent taste, but also reduces the digestibility of the leaf proteins essential to the caterpillars' development. The amount of leaf tannin increases as the season progresses, and moths have evolved a life-cycle in which the caterpillars concentrate their development in early spring, when the leaves are least toxic. Very young caterpillars are only able to eat the most tender leaves, so their hatching needs to coincide as closely as possible with the breaking of the leaf buds. This strategy can prove hazardous – if the caterpillars hatch too soon they starve, but if they hatch too late they also starve, since the leaves will have become too tough to eat. As a result, the amount of leaf damage caused to individual trees can vary greatly from year to year.

CURRANT GALLS
*(See page 43)*
These currant galls on the male catkins are caused by the gall wasp *Neuroterus quercusbaccarum*. The wasps emerge in early summer.

THE BRANCH IN SPRING
Bud-break and the shedding of ripe pollen from male catkins can vary by as much as one week between individual trees in the same wood. This unpredictability helps to reduce the amount of defoliation caused by caterpillars.

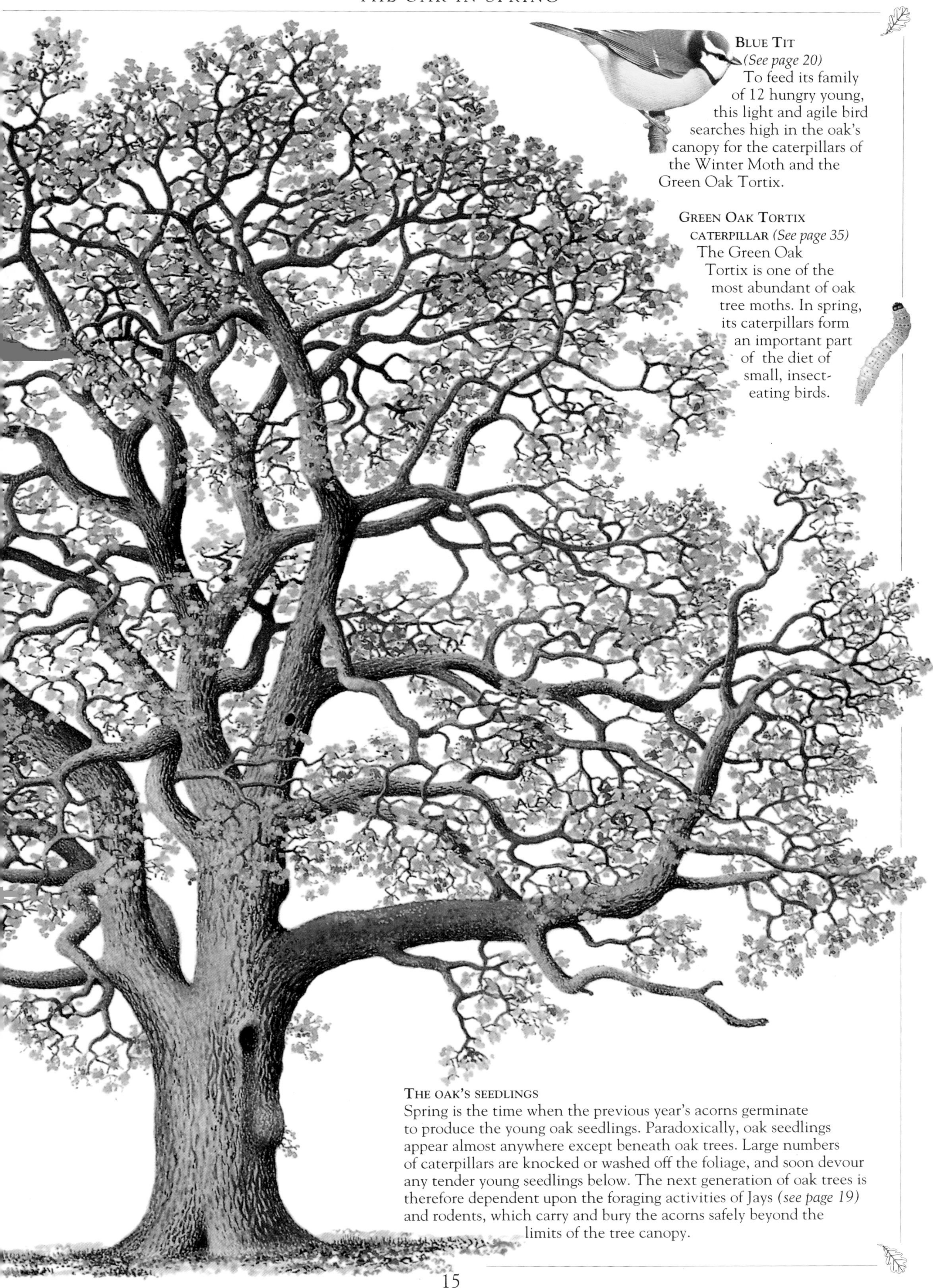

**BLUE TIT**
*(See page 20)*
To feed its family of 12 hungry young, this light and agile bird searches high in the oak's canopy for the caterpillars of the Winter Moth and the Green Oak Tortix.

**GREEN OAK TORTIX CATERPILLAR** *(See page 35)*
The Green Oak Tortix is one of the most abundant of oak tree moths. In spring, its caterpillars form an important part of the diet of small, insect-eating birds.

**THE OAK'S SEEDLINGS**
Spring is the time when the previous year's acorns germinate to produce the young oak seedlings. Paradoxically, oak seedlings appear almost anywhere except beneath oak trees. Large numbers of caterpillars are knocked or washed off the foliage, and soon devour any tender young seedlings below. The next generation of oak trees is therefore dependent upon the foraging activities of Jays *(see page 19)* and rodents, which carry and bury the acorns safely beyond the limits of the tree canopy.

# The Oak in Summer

IN SUMMER, THE OAK TREE supports a huge and diverse community of animals, which exploits every conceivable niche provided by the tree's complex structure. The canopy of an oakwood can be imagined as a vast factory, each acre producing about 6,000 tonnes of roots, wood, and leaves during the year. The green chlorophyll in the leaves absorbs the sun's energy, and converts it into chemical energy by photosynthesis, using water and the carbon dioxide in the atmosphere to produce carbohydrates. The tree's photosynthesis helps to maintain the balance of carbon dioxide in the atmosphere, and also provides the energy upon which the oak's intricate food webs depend. Most members of the oak tree community are not are easily seen. Many of the smaller inhabitants are well camouflaged, hide in cracks and crevices, and "freeze" when disturbed.

**Summer Chafer**
*(See page 39)*
Also called the June bug, the Summer Chafer flies high among the branches.

**Buff-tip Moth caterpillar**
*(See page 30)*
The caterpillar of the Buff-tip Moth feeds in colonies in late summer, separating before pupating in the soil over the winter. The caterpillars also feed on sallow, willow, hazel, elm, and lime.

**The branch in summer**
The leaves of the second crop of shoots *(see opposite)*, produced in mid-summer, are larger and paler than those of the first crop, and usually remain more or less undamaged by insects. The whole shoot normally falls in the autumn with the rest of the leaf crop.

**Wood Warbler**
*(See page 21)*
The song of the Wood Warbler is one of the most characteristic sounds of the oakwood in early summer. It has been likened to the sound of a small coin spinning on a marble slab.

**The tree in summer**
By the middle of summer, the oak tree is beginning to look decidedly tatty. The feeding activities of the enormous numbers of moth caterpillars, weevils, and chafers are showing their effects, leaving most of the leaves shredded and holed. In some years, the tree may be almost defoliated by mid-summer. A second crop of new, leafy shoots now begins to appear, especially on the young trees. This is called lammas growth, since it appears during Lammastide, the traditional feast of the first fruits of late summer.

# The Oak in Autumn

ACORNS, THE OAK'S AUTUMNAL CROP, are one of the most important food resources for woodland birds and mammals. In medieval Europe, commoners' rights to pasture swine in woods were jealously guarded, because acorns were an essential part of the pigs' diet. The feeding pigs would inadvertently bury some acorns as they churned up the soil, improving these acorns' germination chances. In a good acorn year, or mast year, a single oak may produce up to 90,000 acorns. Wood Pigeons gorge themselves on the acorns while they are still green and on the tree. A single Wood Pigeon can consume about 120 acorns a day, and is able to carry up to 70 acorns in its crop. Grey Squirrels also consume and hoard large numbers of acorns, and both Rooks and Jays enjoy this seasonal feast. Although small mammals such as Wood Mice and Bank Voles are also partial to acorns, they are not responsible for significant acorn losses.

CHERRY GALLS
*(See page 42)*
These cherry galls are caused by the gall wasp *Cynips quercusfolii*. The larva pupates in the gall during the autumn. Only females emerge, departing from the galls after the leaves have fallen in mid-winter.

SPINDLE SHANK FUNGUS
*(See page 50)*
Warm, wet autumn days produce a wondrous crop of toadstools. Each toadstool is connected to the roots of a neighbouring tree via a network of microscopic threads called hyphae. The Spindle Shank Fungus is common in oakwoods, and grows in clumps at the oaks' bases.

THE BRANCH IN AUTUMN
The foliage of all deciduous trees acquires characteristic autumnal tints prior to leaf fall. The yellow-brown of the oak is due to carotenoid pigments, which are revealed when the green chlorophyll in the leaf is lost as the days shorten.

**JAY**
*(See page 22)*
Jays collect large numbers of acorns in the autumn, which they carry away and bury for the winter. A good proportion of these is never recovered, and germinates the following spring as part of the next generation of oak seedlings.

**THE LITTER FALL**
*(See page 58)*
Oaks start to lose their leaves in the middle of autumn, but yellow leaves can still be seen on the trees in a mild winter. The autumn leaf fall produces a blanket of leaves, called litter, which forms a protective bed for germinating acorns and provides shelter and food for invertebrates and small animals. It decomposes to form humus, returning to the soil nutrients that were taken by the tree earlier in the year and feeding millions of soil organisms.

# PERCHING BIRDS

IN SUMMER, 40 HECTARES (100 ACRES) of oakwood can support between 300 and 400 birds of more than 30 different species. Most of these are small passerines, or perching birds, no more than about 15 cm (6 in) long. This rich and diverse birdlife relies on the abundance of food and nest sites. Most small woodland birds are insect-eaters, and even seed-eaters such as the Chaffinch feed their young exclusively on invertebrates. An oak tree in late spring supports a huge population of insects at exactly the same time that birds are feeding their hungry broods. The hole-nesting species, such as the Great Tit, Blue Tit, Pied Flycatcher, and Redstart, can choose between the many holes found in old and decaying trees, while other species are able to find nest cover in the tree canopy and in undergrowth.

**CHIFFCHAFF**
***Phylloscopus collybita***
The Chiffchaff and the Willow Warbler are virtually indistinguishable except by their song. Both are summer visitors, but increasing numbers of Chiffchaffs have begun to overwinter in northern Europe. The "chiff, chiff, chaff" song is one of the familiar heralds of spring. The female lays five or six eggs in a dome-shaped nest built in undergrowth close to the ground.

CHIFFCHAFF
*Phylloscopus collybita*

*Legs are darker than those of the Willow Warbler*

FEMALE

*White, bar-shaped patch across wings*

PIED FLYCATCHER
*Muscicapa hypoleuca*

MALE

WILLOW WARBLER
*Phylloscopus trochilus*

**WILLOW WARBLER**
***Phylloscopus trochilus***
This is the most abundant and widespread summer migrant to central and northern Europe. It arrives back in Europe from its winter quarters in West Africa during early spring. The Willow Warbler builds its feather-lined, dome-shaped nest on the ground, concealed among the vegetation in a woodland clearing. It feeds mainly on small flies, moth caterpillars, and other small insects and spiders.

**PIED FLYCATCHER**
***Muscicapa hypoleuca***
Pied Flycatchers are summer visitors that arrive in Europe in spring. They favour woods with little undergrowth and need a plentiful supply of nesting holes. In Britain, Pied Flycatchers are confined almost entirely to western Sessile oakwoods, but the provision of nest boxes has helped to extend their range in recent years.

*Bright blue cap and white cheeks*

**BLUE TIT**
***Parus caeruleus***
The Blue Tit has a distinct liking for oakwoods. It builds its feather-lined nest in a hole in a tree, and towards the end of spring the female lays between seven and thirteen white, speckled eggs, or sometimes even more.

BLUE TIT
*Parus caeruleus*

GREAT TIT
*Parus major*

*Black-and-white head*

**GREAT TIT**
***Parus major***
The Great Tit feeds on the oak's lower branches, and also on the ground. It needs larger nest holes than the Blue Tit. The nest is lined with fur or hair, and the female lays between five and twelve eggs.

**WOOD WARBLER**
***Phylloscopus sibilatrix***
Much more exclusively a woodland bird than either the Chiffchaff or the Willow Warbler, the Wood Warbler favours mature woodland with few shrubs and little ground vegetation. It is especially characteristic of the Sessile oakwoods of western Britain.

**REDSTART**
***Phoenicurus phoenicurus***
A summer visitor from West Africa, the Redstart arrives in Europe in mid- to late spring. It builds its nest in a hole in a tree, and the female lays six pale blue eggs in late spring. The young are fed almost exclusively on small insects such as sawflies and caterpillars. Redstarts are more plentiful in Sessile than in Common oakwoods.

**TREE CREEPER**
***Certhia familiaris***
Although this tiny bird is common in oakwoods throughout Britain, it is often overlooked due to its retiring habits. The Tree Creeper derives its name from its habits – it "creeps" about trees, probing for insects with its curved bill. It builds its nest behind loose bark or among thick ivy. In other parts of Europe, it is found mainly in upland conifer forests, and in lowland broadleaved woods it is replaced by its close relative, the Short-toed Tree Creeper.

**CHAFFINCH**
***Fringilla coelebs***
The Chaffinch is the most abundant woodland bird. It builds its cup-shaped nest, coated with lichens and spider silk, at the end of spring, often concealing it in a crutch in the oak's branches. The Chaffinch's four or five young are fed exclusively on insects, although outside the breeding season it is a seed-eater.

**BLACKCAP**
***Sylvia atricapilla***
The Blackcap has one of the most beautiful songs of all European birds, rivalling both the Blackbird and the Nightingale. It prefers mature woodland with a good shrub layer. Its nest is built in undergrowth about 1 metre (3 ft) above the ground, and the female usually lays five eggs.

# WOODPECKERS & OTHERS

MANY OF THE MOST STRIKING and colourful woodland birds, such as woodpeckers, tits, the Redstart, and the Nuthatch, are hole-nesting species. Generally speaking, these attractive birds are not among the most gifted songsters. Some birds play an important part in the life-cycle of the oak tree. Oaks depend on birds and mammals to disperse their acorns, ensuring that they germinate at a safe distance from the parent tree. Jays are especially significant in this respect, carrying acorns up to a kilometre from the parent tree and burying them in the open.

MISTLE THRUSH
*Turdus viscivorus*

ADULT JAY IN-FLIGHT

NUTHATCH
*Sitta europaea*

JAY
*Garrulus glandarius*

**MISTLE THRUSH**
***Turdus viscivorus***
The Mistle Thrush is the largest European thrush, with a length of about 27 cm (10½ in), and is found mainly on the edges of woodland. It is sometimes known as the storm cock, a name derived from its habit of singing from the top of the highest tree, in defiance of the worst weather. It begins to nest at the end of winter, and up to three broods a year may be produced. Fruit and berries form a large part of the Mistle Thrush's diet, but despite its name, it is rarely seen feeding on mistletoe in Britain.

**NUTHATCH**
***Sitta europaea***
This delightful tree-climbing bird, no more than about 14 cm (5½ in) long, is characteristic of mature, deciduous woodland. Its penetrating, repetitive call is a familiar sound in oakwoods in early spring. The entrance to its nest hole is plastered with mud to reduce the diameter of the entrance to about 3 cm (1⅛ in). The Nuthatch feeds by running up and down the main branches of trees in search of food but, unlike the woodpeckers, it does not use its tail for support.

**JAY**
***Garrulus glandarius***
The most colourful European member of the crow family, the Jay is about 34 cm (13½ in) in length. It is a typical woodland bird, prefering oakwoods with thick undergrowth. Usually feeding on large insects, fruits, and nuts, the Jay collects and buries acorns, often at considerable distances away from the parent oaks. This activity plays a major part in the spread of oak woodland. The Jay is a rather shy bird, and its presence is usually first revealed by its loud, raucous calls.

**LESSER SPOTTED WOODPECKER**
***Dendrocopos minor***
This tiny, sparrow-sized woodpecker prefers open woodland with a scattering of tall, dead trees. It excavates its nest high up a dead tree, at a height of up to 25 metres (80 ft). The entrance to the nest is often on the underside of a branch. The Lesser Spotted Woodpecker feeds almost exclusively on small insects. It is a shy bird, whose presence may be given away only by its characteristic drumming, which is quieter but more high-pitched than that of the Great Spotted Woodpecker.

**GREAT SPOTTED WOODPECKER**
***Dendrocopos major***
The woodpeckers are beautifully adapted for life as tree-climbing birds. They can run rapidly up and down tree trunks, firmly supported by their stiff, pointed tail feathers, and with their heads ever upwards. Their feet have four toes, two pointing forwards, and two pointing backwards. The hard, chisel-shaped bill is used to drill into soft wood, in search of insects, or to excavate the nest hole. The chiselling bill produces a drumming sound. The long, extensible tongue is used to probe for insects in holes and crevices. Great Spotted Woodpeckers feed on a variety of insects, seeds, eggs, and young birds, depending on the season. They can be found in woodland throughout Europe, and their drumming is a familiar springtime sound. An adult is about 23 cm (9 in) in length.

**GREEN WOODPECKER**
***Picus viridis***
This large woodpecker, with a length of around 32 cm (12½ in), is unmistakable in appearance. It is sometimes better known as the yaffle, a reference to its loud, "laughing" call. Unlike the Lesser Spotted and Great Spotted Woodpeckers, it rarely drums, and feeds mostly on the ground, almost exclusively on ants. Both the male and the female share the excavation of the nest hole, which may be up to 5 metres (17 ft) above the ground.

LESSER SPOTTED WOODPECKER *Dendrocopos minor*

GREAT SPOTTED WOODPECKER *Dendrocopos major*

GREEN WOODPECKER *Picus viridis*

# LARGE WOODLAND BIRDS

THE BIRDS OF PREY occupy the top of the woodland food chain. Some, like the Sparrowhawk, are specialist bird predators. The diet of others, like the Buzzard, is more diverse, consisting of small mammals, rabbits, and young birds. The owls play a similar part at night, feeding largely on small mammals. The Red Kite mixes living prey with carrion, especially dead sheep.

Almost all birds of prey have suffered losses in this century, the Red Kite only just escaping extinction in Britain. Fortunately a more enlightened attitude among landowners, the banning of the organochlorine pesticides, and strict legal protection have resulted in a significant recovery in all woodland birds of prey in recent years. Gamekeepers now appreciate that owls and Pheasant chicks can coexist. Not all large birds are birds of prey. Doves and pigeons are specialized plant feeders, and the Woodcock is a wader relative of the Snipe and Curlew.

ADULT RED KITE IN FLIGHT

RED KITE *Milvus milvus*

**RED KITE**
***Milvus milvus***
This beautiful bird was once widespread in Britain, but persecution drastically reduced its numbers so that by the beginning of this century only a handful of individuals survived in central Wales. In mainland Europe persecution has had a similar catastrophic effect. As a result of rigorous protection, the Welsh population has now increased to more than 50 pairs. Mature Sessile oakwoods in remote valleys are the favourite nesting site. Breeding commences early, the one to three eggs being laid in mid-spring in a nest usually built in the fork of an old oak. Males grow up to 58 cm (23 in) in length, females up to 65 cm (25½ in).

**TAWNY OWL**
***Strix aluco***
The eerie hooting of the Tawny Owl can create a ghostly effect in a darkened oakwood. Rarely seen in the day, at night the Tawny Owl hunts rodents, small birds, and beetles. The most common European owl, it is found everywhere but Ireland and the extreme north. It nests in holes in old trees, but will also readily use nesting boxes. Tawny Owls grow to 38 cm (15 in) in length. Nesting starts early, with two to five eggs being laid in early spring.

TAWNY OWL *Strix aluco*

*Plumage is a finely marked combination of browns, barred with black and chestnut*

WOODCOCK *Scolopax rusticola*

**WOODCOCK**
***Scolopax rusticola***
Found across most of northern and central Europe, the Woodcock is about 36 cm (14½ in) long. It is a solitary and retiring bird, preferring large, damp woods, with a thick undergrowth of brambles, holly, and bracken. If disturbed, it takes off with a rapid, twisting flight, but otherwise it is rarely seen. During the breeding season, the males make territorial dusk flights and utter a characteristic "roding" song, consisting of several deep, croaking notes followed by a high-pitched sneezing sound. Primarily ground-dwelling, the Woodcock probes the forest floor with its long bill for earthworms, larvae, and beetles. It builds a shallow nest lined with dried grass and dead leaves in which to lay its four well-camouflaged, spotted eggs.

ADULT WOOD PIGEON IN FLIGHT

WOOD PIGEON
*Columba palumbus*

ADULT STOCK DOVE IN FLIGHT

STOCK DOVE
*Columba oenas*

BUZZARD
*Buteo buteo*

ADULT BUZZARD IN FLIGHT

PHEASANT
*Phasianus colchicus*

**WOOD PIGEON**
***Columba palumbus***
The Wood Pigeon is so familiar on farmland that it is not commonly thought of as a woodland bird. It measures 40 cm (16 in) in length, and feeds on seeds and herbage, gorging on green acorns in early autumn. Two white eggs are laid on a flimsy platform of twigs, built in the fork of a tree. The pale eggs are easy prey to Jays, Magpies, and squirrels, and both parents brood the eggs so that they are left uncovered as little as possible. The young nestlings, or squabs, are fed on "pigeon milk", a cheesy, protein-rich secretion from the crop in the parents' throats.

**STOCK DOVE**
***Columba oenas***
The Stock Dove is found throughout central and western Europe, although in Britain it has not yet reached the far north. Measuring 34 cm (13½ in) in length, it prefers open woodlands, and nests in holes in old trees, sometimes in loose colonies. Like the Wood Pigeon, the Stock Dove feeds on seeds and leaves on the ground, as well as on the flowers and buds of trees.

**BUZZARD**
***Buteo buteo***
The Buzzard is found across most of Europe except for Ireland and northern Scandinavia. Following a century of persecution, its soaring flight is once again a familiar sight over much of the British countryside. It is an impressive spectacle: Buzzards grow to about 52 cm (21 in) in length, the females being 5 cm (2 in) longer than the males. The Buzzard hunts rabbits and small rodents, together with birds like young Wood Pigeons, and even the occasional Adder. Its nest is built in a fork high in a tree, usually an oak or a pine, where concealment can be combined with a wide field of view. Each pair of Buzzards will have several nests in its territory, occupying a different one each year. Two or three eggs are laid in spring.

**PHEASANT**
***Phasianus colchicus***
The Pheasant was introduced to Europe from central Asia in antiquity, reaching Italy during Roman times and Britain during the 11th century. It is one of the largest game birds, with the females measuring about 58 cm (23 in) in length and the males, with their showy tails, growing to 84 cm (33 in). On sporting estates, numbers are kept up by breeding in pens and feeding over the winter. The Pheasant is mainly a ground-dwelling bird. It scratches and digs for seeds, grain, roots, and shoots, as well as small invertebrates, and roosts in trees at night. It builds a nest hidden in thick vegetation on the ground, and lays eight to fifteen olive-brown eggs.

# MAMMALS

ALTHOUGH MAMMALS are abundant in woodlands they are rarely seen. The exception is the introduced Grey Squirrel, which often reveals its presence by a characteristic "chattering" accompanied by much tail-flicking. Wood Mice are strictly nocturnal, but bats usually emerge before dark. Bank Voles scuttle along their subterranean burrows or are hidden in undergrowth. Badgers are more easily located by their setts, well-trodden paths, and characteristic latrines. Foxes have a characteristic scent, and both Badgers and Foxes call at night. Footprints, or slots, in the soft mud of a stream crossing, or the characteristic droppings, fewmets, are often the first hint of the presence of deer. Shrews and Hedgehogs are less easy to detect, but Moles throw up their familiar mounds.

**PIPISTRELLE BAT**
***Pipistrellus pipistrellus***
The Pipistrelle is one of Europe's most common bats, and is found as far north as Scandinavia. With a wingspan of about 20 cm (8 in), it is also the smallest. During the summer, colonies roost by day in hollow trees, behind loose bark, or among ivy, as well as in buildings.

**WOOD MOUSE**
***Apodemus sylvaticus***
Also known as the Long-tailed Field Mouse, this little rodent occurs throughout Europe as far north as southern Scandinavia. Strictly nocturnal, it can jump, climb, and swim well. Wood Mice feed on seeds, nuts, insects, and roots, and make a nest of grass and moss, sometimes using an old bird's nest. A female can produce six litters of up to nine young each year. An adult is about 7–10 cm (3–4¼ in) in length.

**NOCTULE BAT**
***Nyctalus noctula***
The Noctule is a true woodland dweller, roosting and hibernating in large colonies in holes in trees. A large, robust bat with pointed wings, it can often be seen in the early evening, flying fast and high above the trees hunting chafers, moths, and other large insects. It has a wingspan of 30–40 cm (12–15 in).

**YELLOW-NECKED MOUSE**
***Apodemus flavicollis***
Slightly larger than the Wood Mouse, this mouse is recognizable by the yellow band on its chest, and the dark fur on its upper body. It feeds on seeds, fruit, nuts, and berries, storing surplus food in underground larders during the winter, and often entering houses in cold weather.

**BADGER**
***Meles meles***
Found throughout Europe and Asia, the Badger is mostly nocturnal. It emerges at dusk to feed chiefly on earthworms, but also on nuts, insects, plants, and carrion. Its den, or sett, has several entrances and is used by successive generations, sometimes for centuries. The cubs are born in late winter and appear above ground in early spring. A male, or boar, will grow to about 90 cm (3 ft) long.

GREY SQUIRREL
***Sciurus carolinensis***
The Grey Squirrel was introduced to England from North America at the end of the 19th century. It is found over most of Britain, but it has not reached mainland Europe. It feeds on buds, young shoots, nuts, berries, insects, and eggs. The domed nest, or drey, is built of leaves and twigs high in the branches of a tree. An adult measures about 25 cm (10 in) long, with a 20-cm (8-in) tail.

RED SQUIRREL
***Sciurus vulgaris***
Common throughout Europe, in Britain the Red Squirrel has disappeared from many of its former haunts and is chiefly an animal of pine woodlands. Its decline has been linked to epidemic disease and the spread of the Grey Squirrel. Completely diurnal, it feeds on nuts, berries, bark, insects, and small birds, and stores food for the winter, since it does not hibernate. It is about 20 cm (8 in) long, with an 18-cm (7-in) tail.

BANK VOLE
***Clethrionomys glareolus***
Common across Europe and central Asia, the Bank Vole is very much an animal of the woodland edge. Its diet is more vegetarian than that of the Wood Mouse, and its diurnal habits lead it to favour a denser vegetation. A complex burrow system is made close to the surface, and territory is maintained by marking the boundaries with urine. Several litters are raised each year in a ball-shaped underground nest of grass and leaves. The vole grows to about 10 cm (4 in) in length.

FALLOW DEER
***Dama dama***
This beautiful animal was once used to stock royal parks and forests for hunting. Fallow Deer have spread spectacularly in the last 70 years, and are widespread both in natural and managed forests and in large parklands. During early autumn, the bucks herd the does into harems, with much tussling between rival males. The bucks and does remain together for the winter, but the bucks are off again in spring, and keep apart while their antlers are regrowing. The fawns are born in early summer. Deer are fond of acorns, and browse on young shoots and leaves during the summer. An adult buck stands about 90 cm (3 ft) at the shoulder.

*All butterflies shown life sized unless otherwise specified*

# WOODLAND BUTTERFLIES

OAK WOODLANDS ARE THE HOME of some of the rarest and most spectacular butterflies, although only one British species depends solely upon the oak for the caterpillar food plant. In the last 40 years, several species of woodland butterfly have shown a catastrophic decline, coinciding with the abandonment of traditional coppice management (*see page* 8). Chief among the declining species are the High Brown, Pearl-bordered, and Small Pearl-bordered Fritillaries. Butterfly populations are heavily dependent upon woodland management. The same changes that caused the decrease in the fritillaries have favoured other species more tolerant of the shade of mature woodland. The Speckled Wood began to spread in the 1920s and can now be found in many woodlands. More recently, the White Admiral has been showing signs of a similar exciting recovery.

MALE

WHITE ADMIRAL
*Lodoga camilla*

MALE *obliterae*
COLOUR VARIATION

### WHITE ADMIRAL
### *Lodoga camilla*

This elegant butterfly has lovely markings on the undersides of its wings. The adults appear in early summer, and have a preference for shady woodlands. They bask in the canopy, search for honeydew, or feed on bramble and other flowers in sunny spots close to the ground. The female lays each of her eggs on the tip of a shaded honeysuckle leaf. The caterpillar overwinters within a hibernaculum, which it forms by securing a leaf to a stem with silk to prevent it from being shed, and then folding the leaf down the middle. The White Admiral is found throughout central Europe. In Britain it is confined to southern woodlands, but has been spreading north in recent years.

FEMALE

PURPLE EMPEROR
*Apatura iris*

MALE

### PURPLE EMPEROR
### *Apatura iris*

This magnificent butterfly was named the Emperor of the Woods by 18th-century naturalists. It is restricted in its distribution, but can be found in woods in southern Britain and northern Europe. Adult Purple Emperors emerge in mid-summer, and usually fly high in the canopy, the males establishing territories on master oaks – prominent trees that are occupied year after year. The butterflies feed on aphid honeydew, and are attracted to the sap oozing from tree wounds. Male Purple Emperors are also partial to dung and rotting carrion, a taste exploited by butterfly collectors in the 19th century. The females lay their eggs on the leaves of the Goat Willow, *Salix caprea*, choosing large bushes in prominent, partly shaded conditions. The caterpillars are green, and turn brown in autumn before hibernating attached to a twig.

MALE
PURPLE EMPEROR
(SHOWN X 1¼)

MALE
WHITE ADMIRAL
(SHOWN X 1¼)

*Undersides of both sexes are similar*

**SPECKLED WOOD**
***Parage aegeria***
This pretty butterfly is now common in woodlands throughout most of lowland Europe. It is on the wing from early spring, when the males can be seen patrolling woodland glades, or basking in sunspots in dappled shade. The eggs are laid in warm spots at the edges of clearings, and the caterpillars feed on a variety of grasses.

**SILVER-WASHED FRITILLARY**
***Argynnis paphia***
This spectactular fritillary is now mainly a butterfly of oak woodland. The adults emerge in summer, and feed on bramble and other flowers. The female lays her eggs in the crevices of bark, close to patches of violets, the caterpillars' food plant. The butterfly occurs throughout most of Europe, and its range extends eastwards as far as China.

**PURPLE HAIRSTREAK**
***Quercusia quercus***
The Purple Hairstreak is the only British butterfly that is dependent upon the oak for its caterpillars' food. Colonies seem to be present in almost every lowland oakwood, sometimes in large numbers. The adults emerge in summer. During the day they rest in sunny spots or search for honeydew high in the canopy, but they are very active in the early evening. The eggs are usually laid at the base of a flower bud, and the caterpillars emerge at night to feed on the young leaves. The caterpillars pupate on the soil, where their chances of survival are increased if they are taken into the protection of an ants' nest. The Purple Hairstreak occurs throughout Europe and eastwards to Asia.

*All moths shown life sized unless otherwise specified*

# MACRO MOTHS

THE LARVAE (CATERPILLARS) of more than 200 species of moth feed on the oak, more than on any other European tree. Most of these live on a wide range of other trees and shrubs, but others, such as the Great Prominent, are found only on the oak. The numbers of caterpillars on a single oak tree in late spring and early summer are astronomical, and represent an essential source of food for many small birds such as tits, warblers, and flycatchers, just when they are feeding their young. Most moths are inactive by day, and the dull colour of their forewings ensures that they are well camouflaged against the trunks and branches on which they rest. Others, like the Merveille du Jour and the Black Arches, have evolved patterns resembling lichens, and a few of them are green.

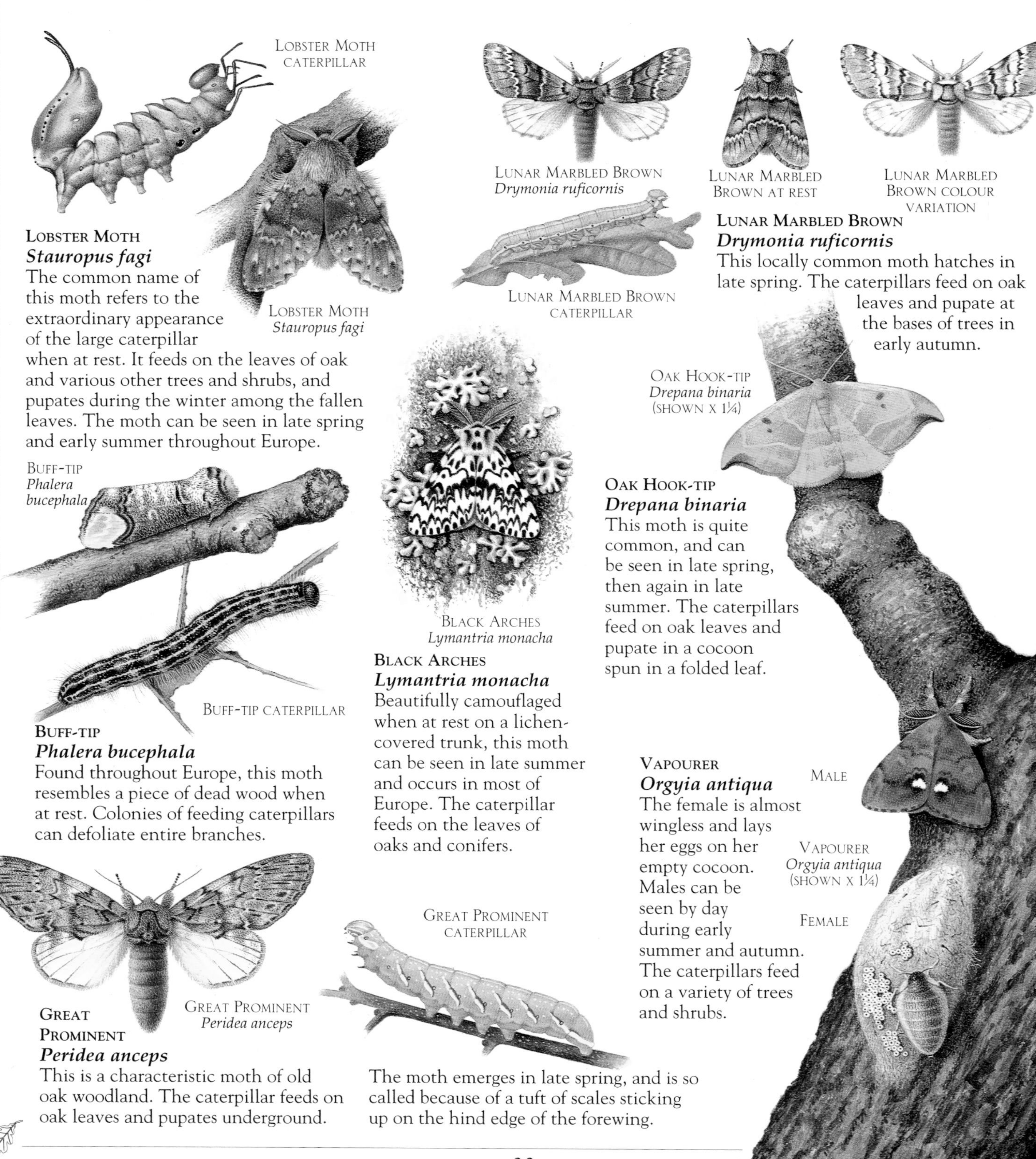


**LOBSTER MOTH**
***Stauropus fagi***
The common name of this moth refers to the extraordinary appearance of the large caterpillar when at rest. It feeds on the leaves of oak and various other trees and shrubs, and pupates during the winter among the fallen leaves. The moth can be seen in late spring and early summer throughout Europe.

**LUNAR MARBLED BROWN**
***Drymonia ruficornis***
This locally common moth hatches in late spring. The caterpillars feed on oak leaves and pupate at the bases of trees in early autumn.

**OAK HOOK-TIP**
***Drepana binaria***
This moth is quite common, and can be seen in late spring, then again in late summer. The caterpillars feed on oak leaves and pupate in a cocoon spun in a folded leaf.

**BUFF-TIP**
***Phalera bucephala***
Found throughout Europe, this moth resembles a piece of dead wood when at rest. Colonies of feeding caterpillars can defoliate entire branches.

**BLACK ARCHES**
***Lymantria monacha***
Beautifully camouflaged when at rest on a lichen-covered trunk, this moth can be seen in late summer and occurs in most of Europe. The caterpillar feeds on the leaves of oaks and conifers.

**VAPOURER**
***Orgyia antiqua***
The female is almost wingless and lays her eggs on her empty cocoon. Males can be seen by day during early summer and autumn. The caterpillars feed on a variety of trees and shrubs.

**GREAT PROMINENT**
***Peridea anceps***
This is a characteristic moth of old oak woodland. The caterpillar feeds on oak leaves and pupates underground. The moth emerges in late spring, and is so called because of a tuft of scales sticking up on the hind edge of the forewing.

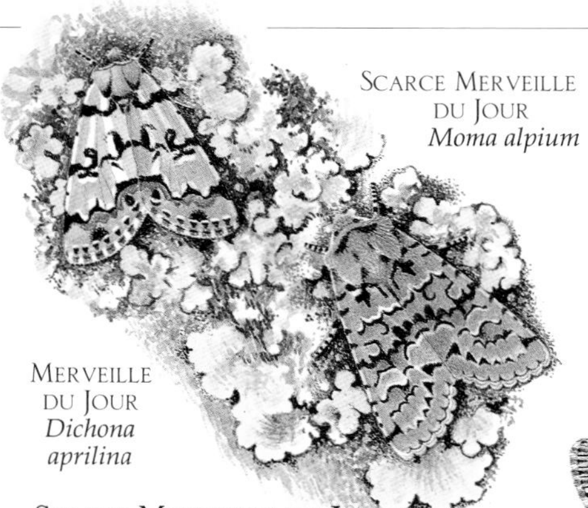

SCARCE MERVEILLE DU JOUR *Moma alpium*

MERVEILLE DU JOUR *Dichona aprilina*

**SPRAWLER**
***Brachionycha sphinx***
Widely distributed in Europe's deciduous woodlands, the Sprawler emerges from its cocoon in autumn to lay its eggs in crevices in bark. When disturbed, the caterpillar adopts the characteristic position, with head and thorax raised, that gives the moth its name.

COPPER UNDERWING *Amphipyra pyramidea*

COPPER UNDERWING CATERPILLAR

SPRAWLER *Brachionycha sphinx*

SPRAWLER CATERPILLAR

**SCARCE MERVEILLE DU JOUR**
***Moma alpium***
**MERVEILLE DU JOUR**
***Dichona aprilina***
These inhabitants of old oakwoods are wonderfully camouflaged on lichen-covered trunks. The caterpillars of both species feed on oak leaves and buds.

**COPPER UNDERWING**
***Amphipyra pyramidea***
This is a common species of oak woodland, seen from late summer to mid-autumn. The caterpillar is found in late spring, most often on oak trees, and pupates below ground.

LIGHT CRIMSON UNDERWING *Catocala promissa*

**LIGHT CRIMSON UNDERWING**
***Catocala promissa***
This species is found in southern and central Europe, but is quite rare in Britain. Despite its striking-coloured underwings, it is well camouflaged when resting on a lichen-covered trunk. The caterpillars hatch in late spring and feed on oak buds, then pupate among lichens in early summer. The adults can be seen in mid- to late summer.

DARK CRIMSON UNDERWING, LARGER AND DARKER THAN C. *Promissa*

WELL-CAMOUFLAGED LIGHT CRIMSON UNDERWING (SHOWN X 2)

DUN-BAR *Cosmia trapezina*

DARK VARIATION LIGHT VARIATION

**DUN-BAR**
***Cosmia trapezina***
This moth hatches in late summer, and is common in deciduous woodland. The caterpillars feed on leaves but are notorious for also eating other larvae, even of their own species.

SATELLITE *Eupsilia transversa*

SATELLITE CATERPILLAR

**GREEN SILVER-LINES** ***Pseudoips fagana***
**SCARCE SILVER-LINES** ***Bena prasinana***
These two moths are well camouflaged among foliage during the day. The common Green Silver-lines appears in early summer in deciduous woodland and hedgerows across Europe. The Scarce Silver-lines also emerges from its cocoon in summer and is found mainly in southern and central Europe.

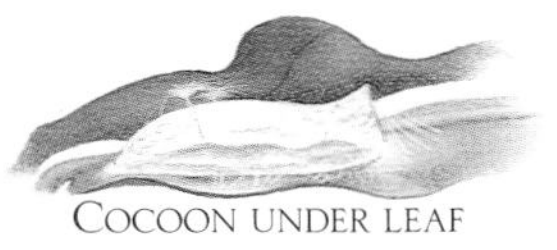

COCOON UNDER LEAF

OAK NYCTEOLINE *Nycteola revayana*

**OAK NYCTEOLINE**
***Nycteola revayana***
Most common in southern regions, this moth emerges in summer and overwinters among ivy or yew. The caterpillar pupates in a cocoon on the underside of a leaf.

**SATELLITE**
***Eupsilia transversa***
Common in mainland Europe and southern Britain, the Satellite emerges in autumn to feed on ivy blossom and survives until the following spring. The young caterpillars feed on the leaves of trees and shrubs, but later turn carnivorous, feeding on larvae and aphids, before pupating in a cocoon below ground.

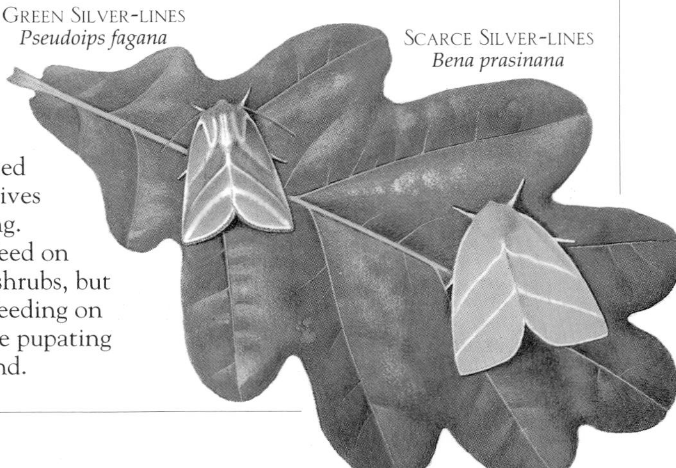

GREEN SILVER-LINES *Pseudoips fagana*

SCARCE SILVER-LINES *Bena prasinana*

*All moths shown life sized unless otherwise specified*

# GEOMETRID MOTHS

MOTHS HAVE EVOLVED excellent camouflage coloration to protect resting adults during the day, and have also evolved successful strategies to protect caterpillars. Caterpillars of moths belonging to the Geometridae family possess only two pairs of abdominal legs at the rears of their bodies. They progress by a series of loops, giving them the popular name of loopers. Many bear a remarkable resemblance to twigs, and "freeze" when disturbed by predators, holding their bodies outstretched and rigid. Several moths of this family, such as the Winter Moth and Mottled Umber, have life-cycles in which the adults emerge in mid-winter. The females are often wingless, and after pupating in the soil, creep up tree trunks to mate, and lay their eggs close to young leaf buds.

MAIDEN'S BLUSH CATERPILLAR

MAIDEN'S BLUSH *Cyclophora punctaria*

FALSE MOCHA *Cyclophora porata*

**MAIDEN'S BLUSH**
***Cyclophora punctaria***
The Maiden's Blush is widespread in oak woodlands. It flies during mid-summer, and sometimes produces a second generation at the end of summer. Its caterpillars feed exclusively on oak.

**FALSE MOCHA**
***Cyclophora porata***
The False Mocha resembles the Maiden's Blush, but has ringed dots on its wings and less distinct crosslines. It flies at the beginning of summer, and sometimes produces a second generation at the end of summer. The caterpillar feeds on oak and birch.

**PEPPERED MOTH**
***Biston betularia***
The pale colour, "peppered" with black, gives the typical form of this moth perfect protection on lichen-covered tree trunks. In 1848, a black form, the variety *carbonaria*, was found in Manchester. This "melanic" form is also found in mainland Europe, and in some areas has almost completely replaced the typical form. The black form is especially predominant in heavily industrialized areas where atmospheric pollution has killed lichens and coated tree trunks with a layer of dust. Against this polluted background, the typical form would be hopelessly conspicuous.

SCORCHED WING *Plagodis dolabraria*

**SCORCHED WING**
***Plagodis dolabraria***
The Scorched Wing is widespread in oak woodlands throughout Europe. It flies in early summer, and the caterpillar feeds on oak and other trees in late summer.

LUNAR THORN *Selenia lunularia*

AUGUST THORN *Ennomos quercinaria*

**THORN MOTHS**
The thorns can be recognized by the characteristic jagged edges to their wings. Both the August Thorn and the Lunar Thorn fly during late summer, and their twiglike caterpillars feed on oak and other trees.

EGG (SHOWN X 20)

PUPA (SHOWN X 1½)

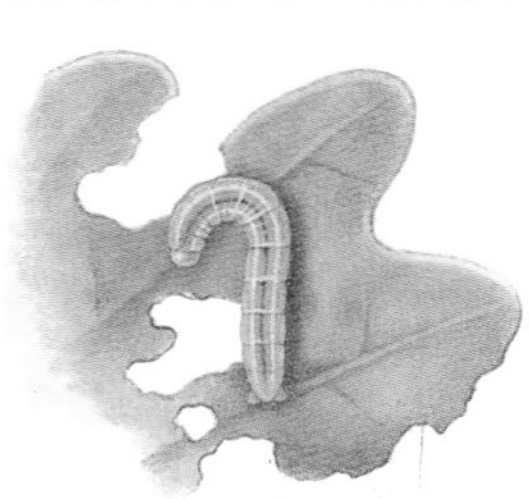

CATERPILLAR (SHOWN X 1½)

FEMALE

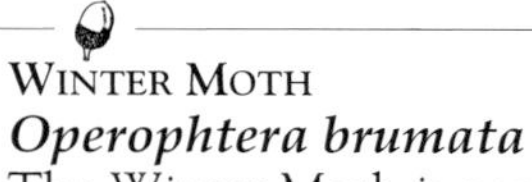

**WINTER MOTH**
***Operophtera brumata***
The Winter Moth is common throughout the British Isles. As its name implies, it emerges from its underground pupa in winter. The female is wingless and creeps up the trunks of trees at night, where she attracts the normally winged males. After mating, she continues upwards to lay her eggs close to the dormant leaf buds. The caterpillars hatch just as the young leaves are unfolding. When mature, they descend to the ground on silken threads to pupate in the soil. As well as feeding on oak, the caterpillars are found on a range of other trees and are a serious pest of orchards.

MALE

WINTER MOTH *Operophtera brumata* (SHOWN X 1½)

MARBLED PUG
*Eupithecia irriguata*
(SHOWN X 1½)

BRINDLED PUG
*Eupithecia abbreviata*
(SHOWN X 1½)

OAK-TREE PUG
*Eupithecia dodoneata*
(SHOWN X 1½)

**MARBLED PUG**
***Eupithecia irriguata***
Found across Europe, in Britain this moth is generally restricted to the New Forest. The caterpillars feed on oak, and the adults fly in late spring.

**BRINDLED PUG**
***Eupithecia abbreviata***
The Brindled Pug is widespread in oakwoods. The adults emerge in spring and rest during the day on tree trunks.

**OAK-TREE PUG**
***Eupithecia dodoneata***
This moth is found in mainland Europe and southern Britain. The adults emerge in late spring, and caterpillars feed on oak leaves during mid-summer.

SPRING USHER
*Agriopis leucophaearia*

**SPRING USHER**
***Agriopis leucophaearia***
Widespread around the world, the Spring Usher, as its name suggests, emerges at the beginning of the year. In late winter the males can be found resting during the day on tree trunks. The females are wingless, and the caterpillars feed on young oak leaves in spring.

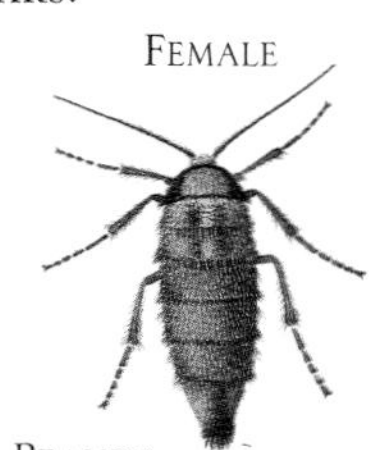

MALE

FEMALE

SMALL BRINDLED BEAUTY
*Apocheima hispidaria*

**SMALL BRINDLED BEAUTY**
***Apocheima hispidaria***
The Small Brindled Beauty is distributed throughout Europe. It emerges in spring, and in common with several other winter or early spring species, the female is wingless. The caterpillars feed on oak leaves in early summer before spinning down to pupate in the soil.

**OAK BEAUTY**
***Biston strataria***
This attractive moth is widely distributed in Europe. In early spring it rests during the day near the bases of tree trunks. The caterpillars feed on oak and other tree leaves during early summer.

OAK BEAUTY
CATERPILLAR

BRINDLED BEAUTY
*Lycia hirtaria*

OAK BEAUTY
*Biston strataria*

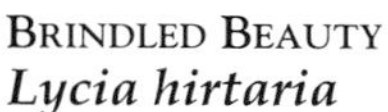

**BRINDLED BEAUTY**
***Lycia hirtaria***
This moth is widely distributed throughout Europe. Its caterpillars can be found in mid-summer, feeding on the foliage of a range of trees including oak, lime, elm, and willow. The adults emerge in spring, and are often found in urban areas, as well as in woodlands.

MOTTLED UMBER
CATERPILLAR

FEMALE
MOTTLED UMBER
*Erannis defoliaria*
(SHOWN X 1½)

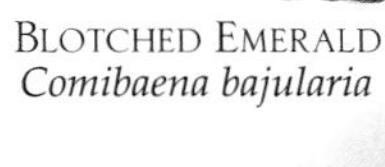

BLOTCHED EMERALD
*Comibaena bajularia*

GREAT OAK BEAUTY
*Boarmia roboraria*

MALE

MALE MOTTLED UMBER
COLOUR VARIATION

**MOTTLED UMBER**
***Erannis defoliaria***
Together with the caterpillars of the Winter Moth and the Green Oak Tortrix, those of the Mottled Umber may sometimes succeed in almost defoliating oaks. The adults emerge in late autumn or mid-winter. The females are wingless, and the males are of widely varying patterns and colours.

**BLOTCHED EMERALD**
***Comibaena bajularia***
The Blotched Emerald is scattered throughout England and central Europe. The caterpillar camouflages itself with fragments of oak leaves and bud scales.

**GREAT OAK BEAUTY**
***Boarmia roboraria***
This attractive moth can be found in the ancient woodlands of southern England and central Europe. The adults emerge in summer, and can be found resting high up on oak trunks. The caterpillars feed during autumn and spring, and hibernate over winter.

# MICRO MOTHS

MOST OF THE MOTHS that feed on oak trees are small "micro" species, with wingspans of less than 25 mm (1 in). Often, two generations emerge in a year. Their caterpillars can leave the foliage in tatters, but the canopy of a tree is a hazardous habitat for small caterpillars. They can be blown, washed, or knocked from their leaves, and are sought by birds with hungry broods to feed. Most species have evolved sophisticated means of protecting themselves. Many of the *Tortrix* moth caterpillars fold or roll leaves around themselves. Others construct portable cases of silk, or spin webs, in order to attach themselves to the undersides of leaves. Several species feed within the tissue of leaves, forming blotch-like or tubular, serpentine mines. These contrivances are peculiar to particular species and so help to identify the moth responsible for them.

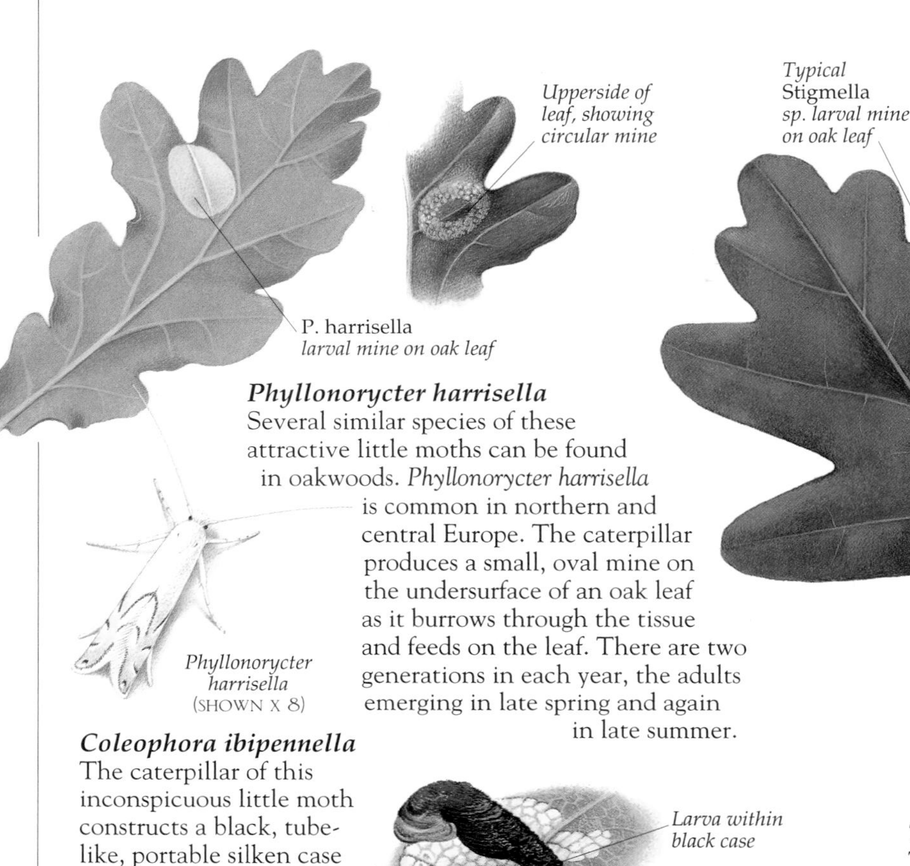

*Upperside of leaf, showing circular mine*

P. harrisella *larval mine on oak leaf*

***Phyllonorycter harrisella***
Several similar species of these attractive little moths can be found in oakwoods. *Phyllonorycter harrisella* is common in northern and central Europe. The caterpillar produces a small, oval mine on the undersurface of an oak leaf as it burrows through the tissue and feeds on the leaf. There are two generations in each year, the adults emerging in late spring and again in late summer.

*Phyllonorycter harrisella* (SHOWN X 8)

*Typical* Stigmella *sp. larval mine on oak leaf*

*Stigmella roborella* (SHOWN X 8)

***Stigmella roborella***
Several species of *Stigmella* are associated with oaks, and *Stigmella roborella* is common throughout western Europe. The caterpillars construct characteristic mines in the leaves, within which they feed, relatively safe from predators. The moths, which emerge in early summer and again in late summer, rest by day on tree trunks.

*Tortrichodes alternella* (SHOWN X 3)

***Coleophora ibipennella***
The caterpillar of this inconspicuous little moth constructs a black, tube-like, portable silken case with two side flaps on the underside of an oak leaf. It lives within this case while feeding. The moth, which flies during summer, is found throughout England and central Europe.

*Larva within black case*

*Coleophora ibipennella* (SHOWN X 4)

***Tortrichodes alternella***
This rather dull-coloured moth is common in oakwoods throughout Britain, and is also found in most of Europe except the south east. The adult moth emerges in early spring, and the caterpillars feed on oak leaves during the summer.

*Scoparia basistrigalis* (SHOWN X 2½)

***Scoparia basistrigalis***
This moth can be found in summer resting head upwards on the trunk of an oak tree, well camouflaged by the colour of its forewings. Found across Europe, in Britain it is generally confined to southern England, and particularly the New Forest.

*Acrocerops brongniardella* (SHOWN X 5)

***Acrocerops brongniardella***
Very common throughout Europe, in Britain this moth is restricted to central and southern England. The moth emerges in late summer, preferring open woodland with young trees. The caterpillars mine in the upper surfaces of oak leaves.

***Eriocrania subpurpurella***
This is one of the most common oak woodland moths, occuring throughout Europe. It flies on sunny days in spring, and the caterpillar feeds in a large, blotch-like mine.

*Eriocrania subpurpurella* (SHOWN X 4)

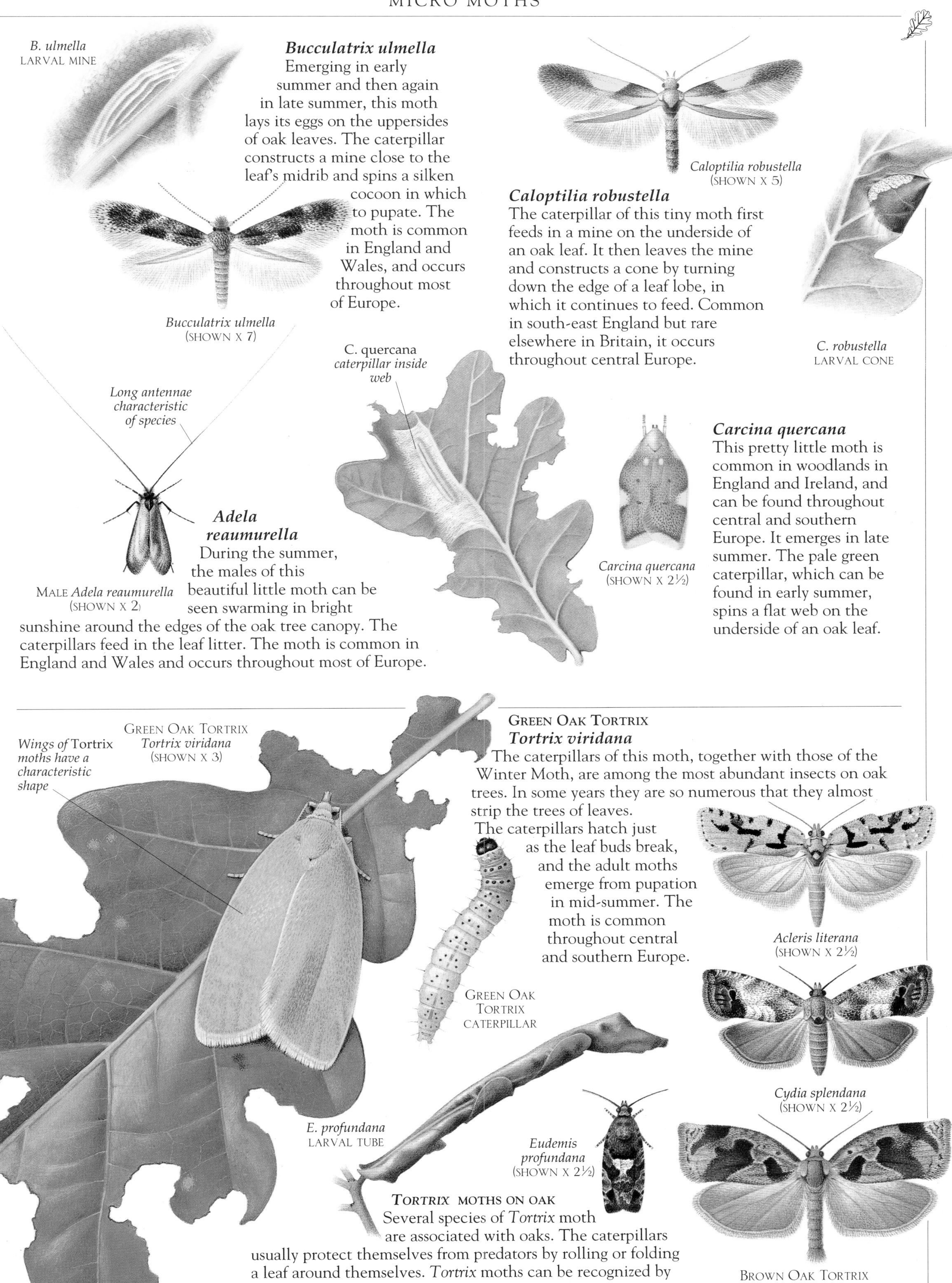


### *Bucculatrix ulmella*

Emerging in early summer and then again in late summer, this moth lays its eggs on the uppersides of oak leaves. The caterpillar constructs a mine close to the leaf's midrib and spins a silken cocoon in which to pupate. The moth is common in England and Wales, and occurs throughout most of Europe.

### *Caloptilia robustella*

The caterpillar of this tiny moth first feeds in a mine on the underside of an oak leaf. It then leaves the mine and constructs a cone by turning down the edge of a leaf lobe, in which it continues to feed. Common in south-east England but rare elsewhere in Britain, it occurs throughout central Europe.

### *Carcina quercana*

This pretty little moth is common in woodlands in England and Ireland, and can be found throughout central and southern Europe. It emerges in late summer. The pale green caterpillar, which can be found in early summer, spins a flat web on the underside of an oak leaf.

### *Adela reaumurella*

During the summer, the males of this beautiful little moth can be seen swarming in bright sunshine around the edges of the oak tree canopy. The caterpillars feed in the leaf litter. The moth is common in England and Wales and occurs throughout most of Europe.

### GREEN OAK TORTRIX
### *Tortrix viridana*

The caterpillars of this moth, together with those of the Winter Moth, are among the most abundant insects on oak trees. In some years they are so numerous that they almost strip the trees of leaves.

The caterpillars hatch just as the leaf buds break, and the adult moths emerge from pupation in mid-summer. The moth is common throughout central and southern Europe.

### *TORTRIX* MOTHS ON OAK

Several species of *Tortrix* moth are associated with oaks. The caterpillars usually protect themselves from predators by rolling or folding a leaf around themselves. *Tortrix* moths can be recognized by the characteristic shape of their forewings.

# BEETLES 1

BEETLES ARE THE MOST numerous group of animals on Earth. They belong to the Coleoptera order of insects and in Britain alone, there are more than 3,700 species. They can be recognized by their hard wing cases, the elytra, which usually cover the abdomen completely. Beetles and bugs are often mistaken for each other, but can be distinguished by their mouthparts – those of beetles are designed for biting, whereas bugs have piercing mouthparts. Many beetles, such as ground beetles and ladybirds, are carnivorous, but others, such as weevils and leafbeetles, are herbivorous. In woodlands, particularly oakwoods, many beetle larvae feed on decaying and dead wood. Their holes and galleries can be revealed if a piece of loose bark is peeled off an old oak stump *(see also page 38)*. Some species of beetle are restricted to old woodland such as ancient royal forests.

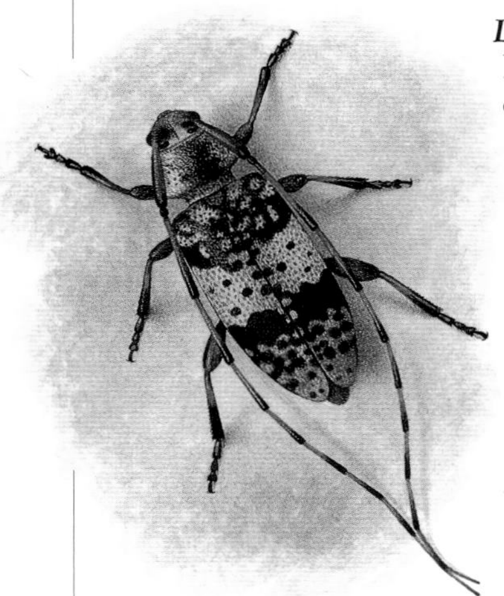

*Leiopous nebulosus*
(SHOWN X 6)

***Leiopus nebulosus***
Found on a range of deciduous trees, but especially on oak, this is one of the longhorn beetles, or Cerambycidae. These beetles derive their name from their long, often threadlike, antennae. The larval stage usually lasts between two and three years. The larvae feed on wood, and prefer decaying timber.

***Calosoma inquisitor***
*Calosoma inquisitor* is a ground beetle belonging to the Carabidae family. Ground beetles have powerful jaws, large eyes, and long legs. They live in the ground, only emerging at dusk to hunt for prey under the cover of darkness. Both larvae and adults are voracious predators. *C. inquisitor* departs from typical ground beetle behaviour, in that it will climb into oak foliage to search for larvae.

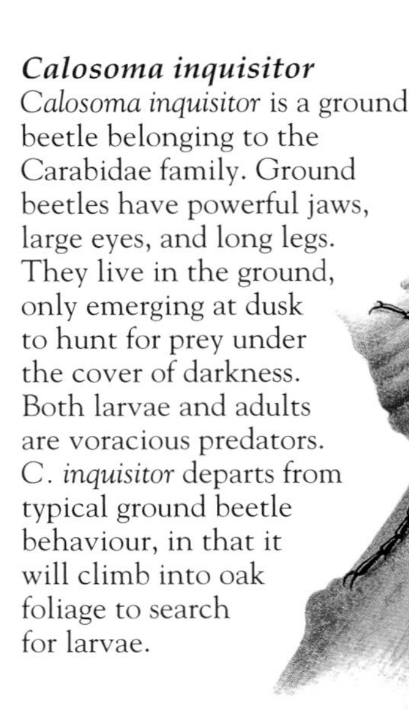

*Calosoma inquisitor*
(SHOWN X 2½)

***Phyllobius argentatus***
Weevils, or Curculionidae, are plant-eating beetles, distinguished by the "snouts" at the fronts of their heads, and by their "elbowed" antennae. *Phyllobius argentatus* can cause considerable damage to young saplings. During the day the adults hide in leaf litter or soil, while at night they feed on oak leaves.

*Phyllobius argentatus*
(SHOWN X 5)

*Oak leaves damaged by weevils*

**LADYBIRDS**

Most ladybirds are carnivorous, and consume an enormous number of aphids. They ensure that their larvae can find food easily by laying their eggs close to groups of aphids. Both larvae and adults keep aphid levels down in gardens or on crops. There are more than 40 different species of ladybird, or Coccinellidae, in Britain. The most familiar is the common Seven-spot Ladybird. Both the Two-spot Ladybird and the Ten-spot have numerous colour varieties. Ladybirds are common in oakwoods throughout Europe, different species being more abundant in different areas.

SEVEN-SPOT LADYBIRD
*Coccinella 7-punctata*
(SHOWN X 2)

TWO-SPOT LADYBIRD
*Adalia bipunctata*
(SHOWN X 3)

*Two-spot Ladybird has black legs*

TWO-SPOT LADYBIRD COLOUR VARIETIES
(SHOWN X 2)

FOURTEEN-SPOT LADYBIRD
*Propylea 14-punctata*
(SHOWN X 3)

TEN-SPOT LADYBIRD
*Adalia 10-punctata*
(SHOWN X 3)

*Ten-spot Ladybird has yellow legs*

TEN-SPOT LADYBIRD COLOUR VARIETIES
(SHOWN X 2)

*Malthinus flaveolus* (SHOWN X 6)

*Malthodes minimus* (SHOWN X 6)

**SOLDIER BEETLES**

Soldier beetles, or Cantharidae, are easily recognized by their soft, parallel-sided wing cases, and long, threadlike antennae. The orange-and-black *Rhagonycha fulva* is one of the many species of soldier beetle that are brightly coloured. Soldier beetles are carnivorous, and feed on small insects. Both *Malthinus flaveolus* and *Malthodes minimus* are common in oak woodlands.

**DEATH-WATCH BEETLE**
***Xestobium rufovillosum***

Many beetle larvae feed on decaying wood, which is enriched with small pieces of fungus. The Death-watch Beetle causes considerable damage to oak timbers in old buildings. Its renowned "ticking" noise, reputed to herald an approaching death, is caused by the beetle's head tapping the wood to attract a mate. In woods, the Death-watch Beetle inhabits decaying oaks.

DEATH-WATCH BEETLE
*Xestobium rufovillosum*
(SHOWN X 4)

*Larval galleries and damage caused by Death-watch Beetles*

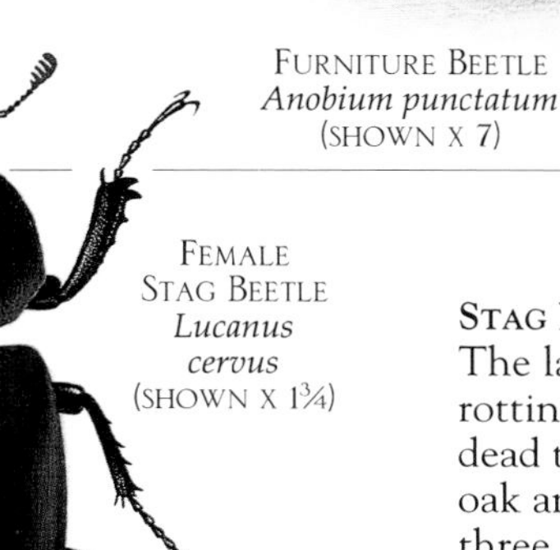

FURNITURE BEETLE
*Anobium punctatum*
(SHOWN X 7)

**FURNITURE BEETLE**
***Anobium punctatum***

The holes of the Furniture Beetle, or "woodworm", are a familiar feature of old timber and furniture. Although most often seen in the home, the Furniture Beetle's original habitat is the forest. The larvae bore a series of irregular tunnels in timber, and pupate just beneath the surface. The adults emerge in mid-summer.

FEMALE STAG BEETLE
*Lucanus cervus*
(SHOWN X 1¾)

**STAG BEETLE LARVA**

The larvae of Stag Beetles feed on rotting wood, and can be found in the dead tree stumps and decaying roots of oak and elm. Stag Beetle larvae take three years to reach maturity, and when fully grown are 12 cm (5 in) long.

STAG BEETLE LARVA

**FEMALE STAG BEETLE**
***Lucanus cervus***

Stag Beetles are most likely to be seen on warm summer evenings, when the males fly in search of mates. Unlike the male, the female Stag Beetle can inflict a painful bite.

*Antlerlike jaws encircle the rival male*

**MALE STAG BEETLE**
***Lucanus cervus***

The Stag Beetle derives its name from the male's enlarged jaws, called mandibles, which resemble the antlers of a stag. When provoked, the Stag Beetle rears up, opening its jaws wide in a ferocious manner. The mandibles are not as formidable as they appear, however, and are used primarily to grasp the female during mating.

*Hard wing cases protect the abdomen and hind wings*

FIGHTING MALE STAG BEETLES
(SHOWN X 1¾)

# BEETLES 2

MANY SPECIES OF BEETLE have evolved highly specialized life-styles to exploit every conceivable niche that the oak tree has to offer. Bark beetles, for example, feed between the bark and the wood, producing characteristic patterns of galleries, and pin-hole borers burrow deep into the wood itself. Some beetles feed directly on wood, producing wood-digesting enzymes in their guts, while other species rely on the more nutritious, fungus-enriched decaying wood. The pin-hole borers have taken specialization one step further by cultivating their own fungi in their bored tunnels, deep within the stem. The larvae of chafers and click beetles feed on plant roots underground, and leaf-rolling weevils protect their larvae within specially constructed leafy tubes.

### OAK BARK BEETLE
***Scolytus intricatus***

As its name suggests, the Oak Bark Beetle is usually confined to oak trees. Its galleries have a distinctive pattern – the "mother gallery", where it lays its eggs, is narrow and horizontal, and the larval galleries are long and vertically arranged.

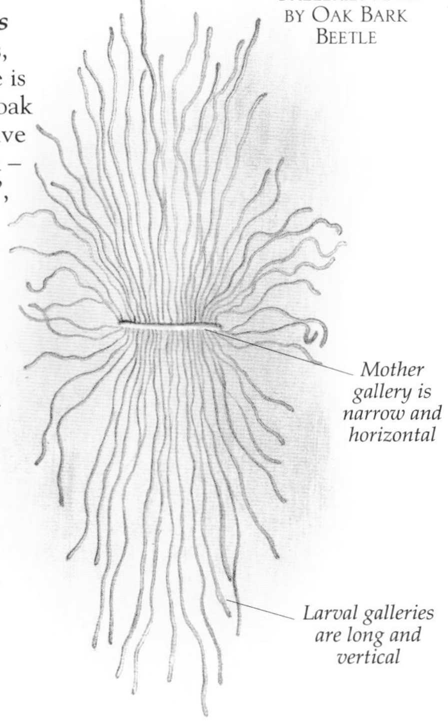

OAK BARK BEETLE *Scolytus intricatus* (SHOWN X 5)

### ELM BARK BEETLE
***Scolytus scolytus***

As its name suggests, the Elm Bark Beetle is usually confined to elm trees, and is notorious as the carrier of the fungus *Ceratostomella ulmi*, the cause of Dutch Elm Disease. The Elm Bark Beetle mines its galleries in the trunks of elms. It bores a vertical "mother gallery", from which all the other galleries radiate. Two generations are produced in a year. The first emerges in spring from larvae that overwintered in the bark, and the second appears in mid-summer.

ELM BARK BEETLE *Scolytus scolytus* (SHOWN X 5)

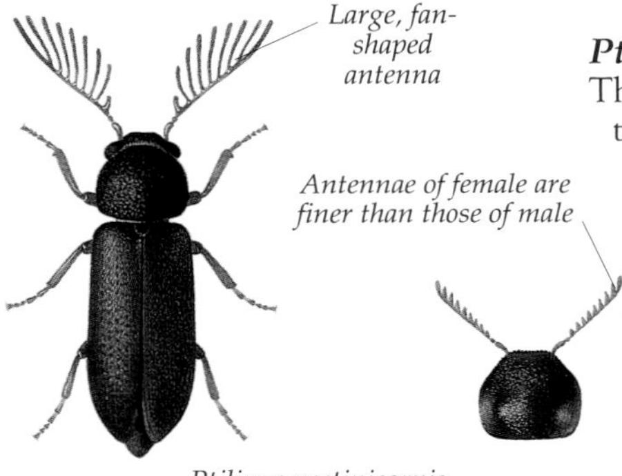


MALE *Ptilinus pectinicornis* (SHOWN X 6) FEMALE

### *Ptilinus pectinicornis*

This tiny beetle is related to the Death Watch Beetle and Furniture Beetle, and is readily distinguished by its characteristic fanned antennae. The beetle is to be found on the trunks of dead trees, in which the larvae feed.

AMBROSIA BEETLE *Trypodendron domesticum* (SHOWN X 8)

### AMBROSIA BEETLE
***Trypodendron domesticum***

Sometimes called pin-hole borers, these minute beetles bore into the wood of living trees, forming characteristic black tunnels. The black coloration is caused by the Ambrosia Fungus, which the beetles cultivate, and on which both adults and larvae feed. The adults lay their eggs in individual side chambers.

### NUT WEEVILS

Nut weevils are easily recognized by their long, thin snouts, or rostrums, which they use to pierce a hole in a nut or plant gall, in which to lay their eggs. *Curculio glandium* and *Curculio venosus* lay their eggs in acorns during early summer, pupating in the soil after the acorns have fallen. *Curculio villosus* and *Curculio pyrrhoceras* lay their eggs in galls.

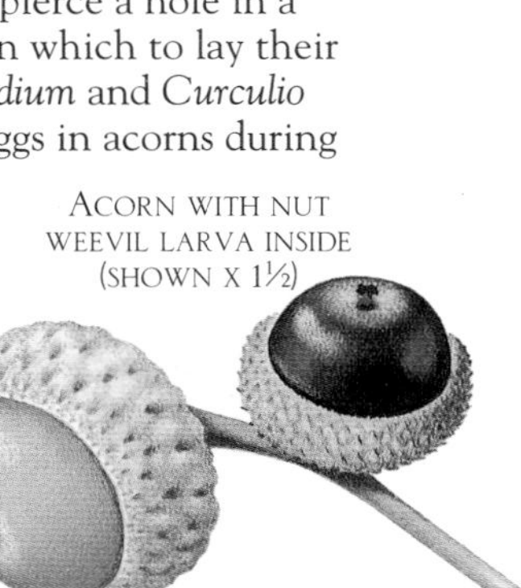

ACORN WITH NUT WEEVIL LARVA INSIDE (SHOWN X 1½)

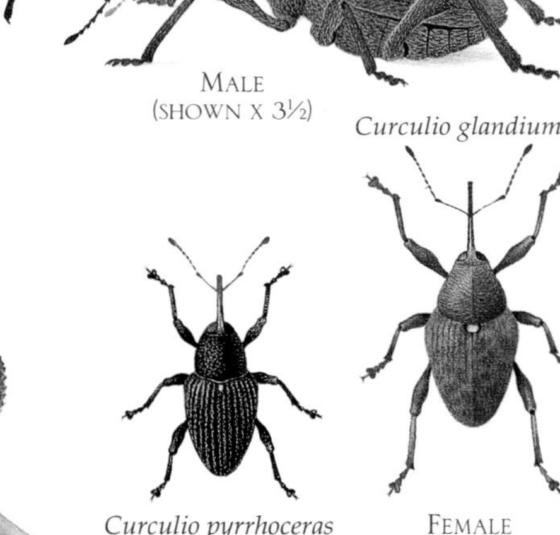

MALE (SHOWN X 3½)

*Curculio glandium*

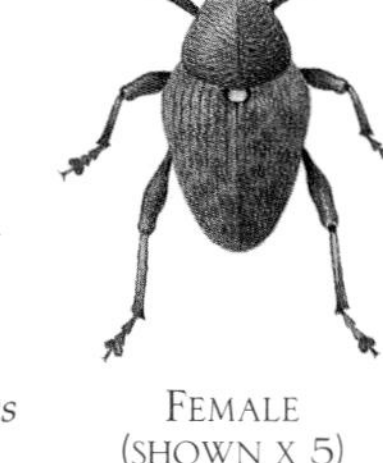

*Curculio pyrrhoceras* (SHOWN X 7)

FEMALE (SHOWN X 5)

*Curculio villosus* (SHOWN X 3½)

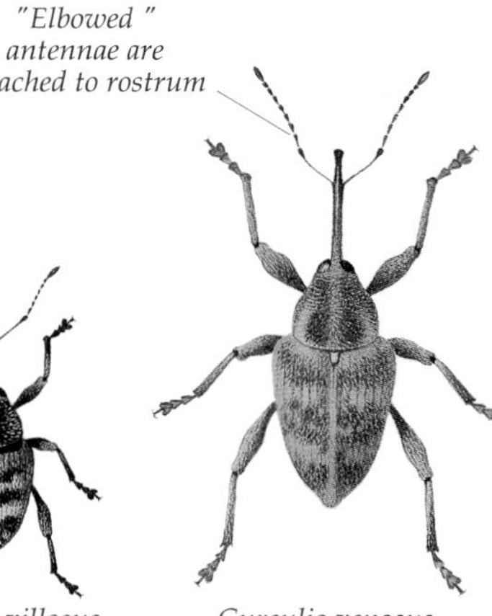


*Curculio venosus* (SHOWN X 3½)

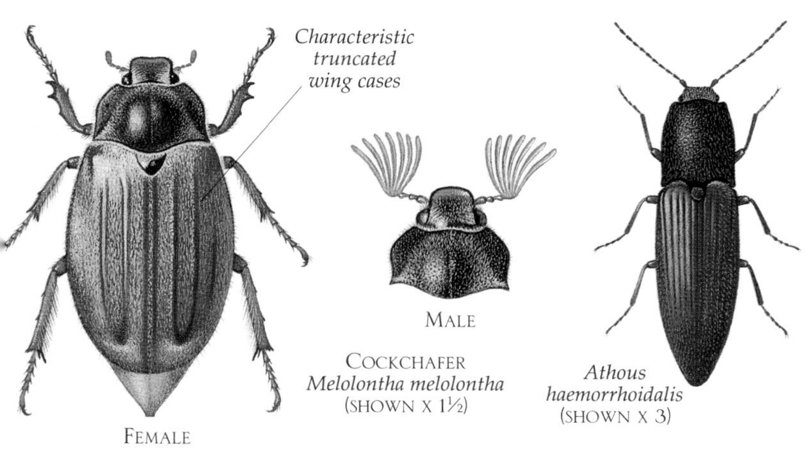


COCKCHAFER
*Melolontha melolontha*
(SHOWN X 1½)

*Athous haemorrhoidalis*
(SHOWN X 3)

### SUMMER CHAFER
### *Amphimallon solstitialis*

Adult Summer Chafers emerge in summer and fly in the evening. They can often be found in large numbers around the tops of trees, where they are eagerly snapped up by large bats. The eggs are laid in the soil, and the larvae feed on the roots of plants for three years before pupating. Like the Cockchafer, the Summer Chafer is one of the scarab beetles. These beetles are distinguished by the characteristic fanlike tips of the antennae, especially in the males.

### COCKCHAFER
### *Melolontha melolontha*

Adult Cockchafers emerge in May, giving them their common name of May-bugs. In some years they emerge in large numbers, and swarm at dusk. The adults feed on oak leaves and can be serious defoliators. The larvae live in the soil, feeding on plant roots.

### *Athous haemorrhoidalis*

Click beetles such as this one are easily recognized by the characteristic shape of their hard, elongated wing cases. Their name refers to their ability to correct themselves when placed on their backs by springing into the air accompanied by an audible "click". Their larvae, commonly known as wireworms, live in the soil and feed on plant roots. *Athous haemorrhoidalis* is a common woodland species.

*Large, delicate hind wings unfold for flight*

SUMMER CHAFER
*Amphimallon solstitialis*
(SHOWN X 2)

RED OAK ROLLER
*Attelabus nitens*
(SHOWN X 4)

*Rolled-up leaf of Red Oak Roller*

### RED OAK ROLLER
### *Attelabus nitens*

This is a distinctive, unmistakable weevil, widespread in oakwoods throughout Europe. The leaf-rolling weevils derive their name from their habit of rolling up a leaf of the host tree into a tube, and laying their eggs inside it. Each species rolls the leaf in a characteristic manner. The Red Oak Roller cuts the leaf to the middle from both sides, folds it along the midrib, and rolls a tube of double thickness. The larvae feed in their tubes for several months before pupating in the soil.

### *Prionus coriarius*

This is one of the largest longhorn beetles, and some individuals reach a length of 4.5 cm (1¾ in). The larvae feed in the stumps and roots of oaks, and pupate beneath the bark. The guts of longhorn beetle larvae contain an enzyme, cellulase, enabling them to digest the wood on which they feed. The adults fly at dusk, hiding by day under bark and in old stumps. When alarmed, the beetle is able to produce a high-pitched sound by rubbing the segments of its thorax together, an activity known as stridulation. *Prionus coriarius* is found throughout most of Europe and North Africa.

*Very long antennae*

*Dendroxena quadrimaculata*
(SHOWN X 3)

### *Dendroxena quadrimaculata*

This distinctive predatory beetle spends its time hunting for moth caterpillars in the foliage of the oak canopy. The larvae, which resemble black, elongated woodlice, live in the soil. *Dendroxena quadrimaculata* is found in oak woodlands in southern England and throughout central Europe.

*Prionus coriarius*
(SHOWN X 2)

*All galls shown life sized unless otherwise specified*

# GALLS 1

THE STRANGE BODIES known as galls are abnormal growths produced by plants in response to a stimulus from another organism, such as an insect, mite, nematode worm, fungus, or bacterium. Over 40 different species produce galls on oaks, more than on any other European plant. By far the largest number are caused by tiny gall wasps from the Cynipidae family *(see also page 42)*. Others are caused by gall midges, moths, and scale insects. The life history of most gall wasps is complicated, exhibiting an alternation between a sexual and an asexual generation in the same year. The sexual generation consists of both males and females, which produce the asexual generation consisting only of females. These females lay eggs that contain the next sexual generation.

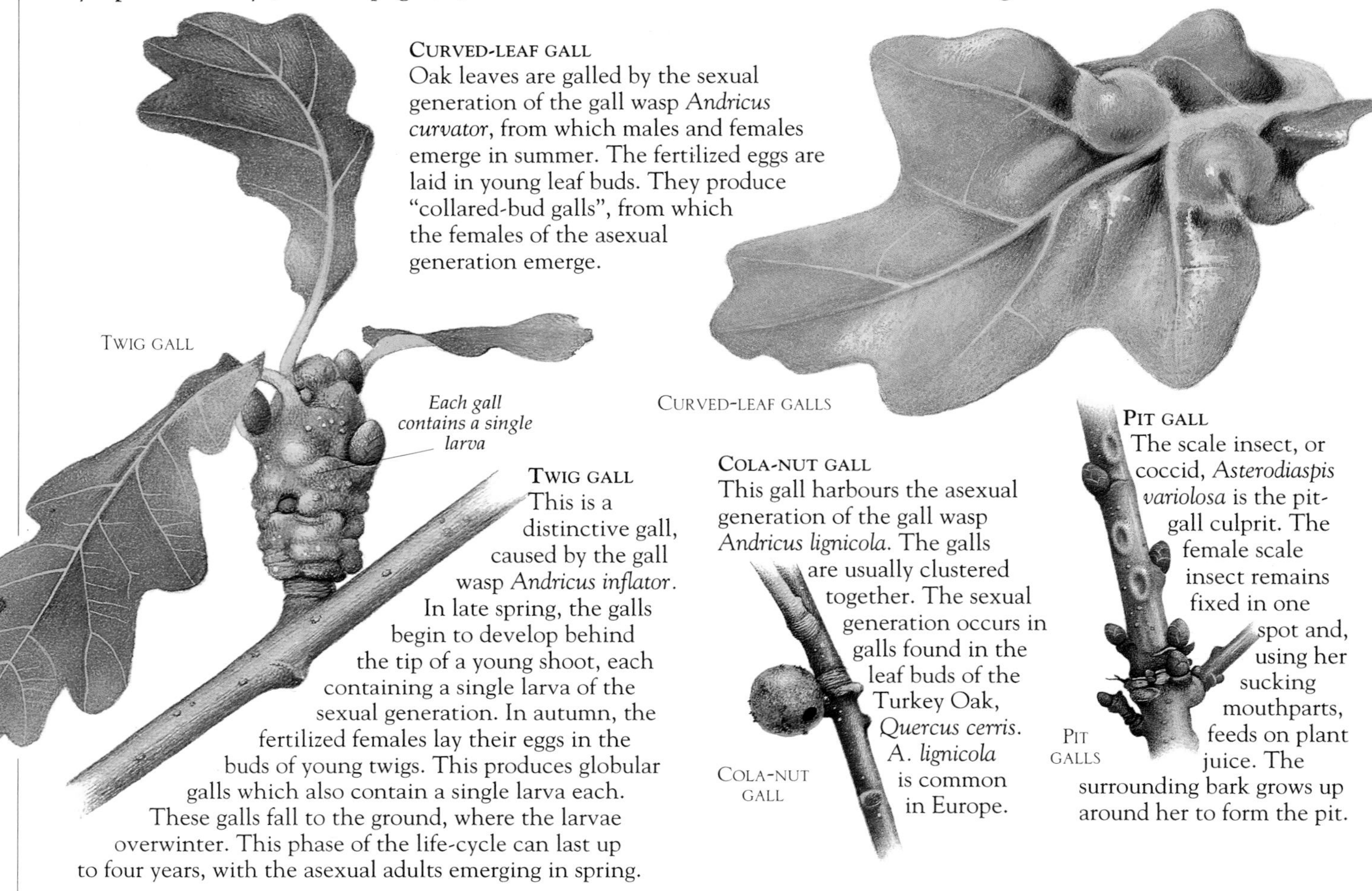


TWIG GALL

CURVED-LEAF GALLS

COLA-NUT GALL

PIT GALLS

**CURVED-LEAF GALL**
Oak leaves are galled by the sexual generation of the gall wasp *Andricus curvator*, from which males and females emerge in summer. The fertilized eggs are laid in young leaf buds. They produce "collared-bud galls", from which the females of the asexual generation emerge.

**TWIG GALL**
This is a distinctive gall, caused by the gall wasp *Andricus inflator*. In late spring, the galls begin to develop behind the tip of a young shoot, each containing a single larva of the sexual generation. In autumn, the fertilized females lay their eggs in the buds of young twigs. This produces globular galls which also contain a single larva each. These galls fall to the ground, where the larvae overwinter. This phase of the life-cycle can last up to four years, with the asexual adults emerging in spring.

**COLA-NUT GALL**
This gall harbours the asexual generation of the gall wasp *Andricus lignicola*. The galls are usually clustered together. The sexual generation occurs in galls found in the leaf buds of the Turkey Oak, *Quercus cerris*. *A. lignicola* is common in Europe.

**PIT GALL**
The scale insect, or coccid, *Asterodiaspis variolosa* is the pit-gall culprit. The female scale insect remains fixed in one spot and, using her sucking mouthparts, feeds on plant juice. The surrounding bark grows up around her to form the pit.

**MARBLE GALL**
Possibly the most familiar of all plant galls, the marble gall is especially common on hedgerow oaks. It is caused by the asexual generation of the gall wasp *Andricus kollari*. The wasps leave the galls in autumn, and the sexual generation produces small galls in the buds of the Turkey Oak. Marble galls contain high levels of tannic acid, and the galls were imported in the 19th century from the Middle East for dyeing and ink-making.

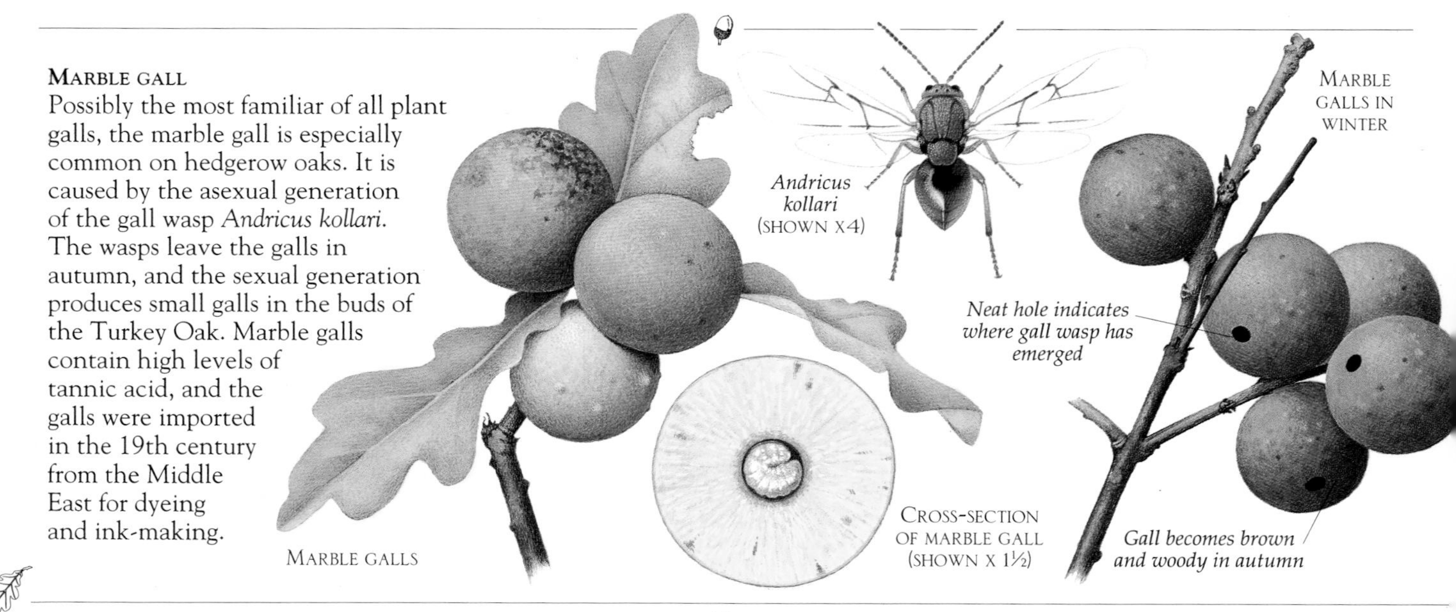


MARBLE GALLS

*Andricus kollari* (SHOWN X4)

CROSS-SECTION OF MARBLE GALL (SHOWN X 1½)

MARBLE GALLS IN WINTER

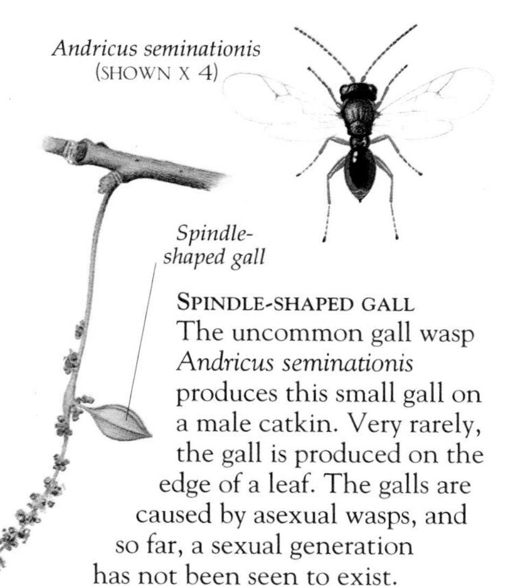


*Andricus seminationis* (SHOWN X 4)

**SPINDLE-SHAPED GALL**
The uncommon gall wasp *Andricus seminationis* produces this small gall on a male catkin. Very rarely, the gall is produced on the edge of a leaf. The galls are caused by asexual wasps, and so far, a sexual generation has not been seen to exist.

**COTTON-WOOL GALL**
This highly conspicuous gall develops on male catkins. As its name suggests, it resembles a ball of cotton wool. It is caused by the uncommon gall wasp *Andricus quercusramuli*. The fluffy mass is around 20 mm (¾ in) in diameter, and conceals up to 20 individual galls fused together. Each gall contains a single larva of the sexual generation. The fertilized eggs are laid in leaf buds, where the pear-shaped galls of the asexual generation develop in late autumn.

COTTON-WOOL GALL

**ARTICHOKE GALL**
This gall is also known as the hop gall, or pineapple gall, but artichoke gall describes its appearance most accurately. It is a distorted and enlarged bud with a hard core, which contains the single larva of the asexual generation of the gall wasp *Andricus fecundator*. The galls are most common on coppiced or hedgerow oaks, and remain on the trees for at least two years. The adult wasps eventually emerge from the galls in spring. The asexual females lay their eggs in male catkins, and the resulting "hairy catkin galls" are small, oval, and hairy. They each contain a single larva of the sexual generation.

OLD ARTICHOKE GALL

ARTICHOKE GALL

**MALPIGHI'S GALL**
This spindle-shaped gall is caused by the gall wasp *Andricus nudus*. It develops in autumn from buds within the leaf axils. The galls fall to the ground, where the larvae overwinter. The asexual females emerge in spring, and lay their eggs in male catkins high in the trees. The resulting "seed galls" are only 2 mm (1⁄16 in) long, and are difficult to find.

MALPIGHI'S GALL (SHOWN X 3)

**KNOPPER GALL**
This extraordinary growth is caused by the asexual generation of the gall wasp *Andricus quercuscalicis*. It is widespread in England and Wales, and for some years it was feared that its spread would severely affect the acorn crop. The sexual generation causes tiny, green galls on the male catkins of the Turkey Oak.

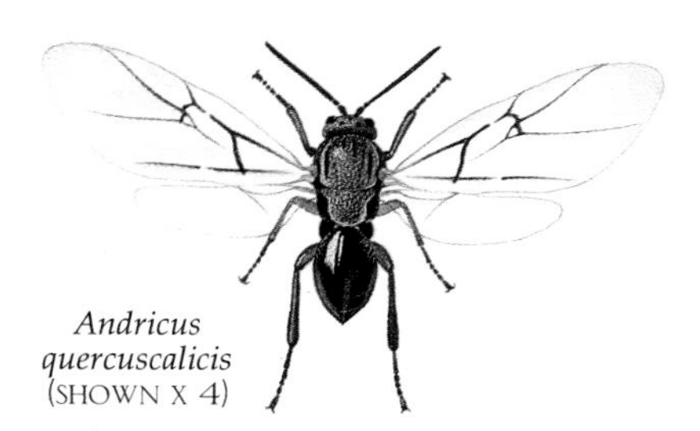

*Andricus quercuscalicis* (SHOWN X 4)

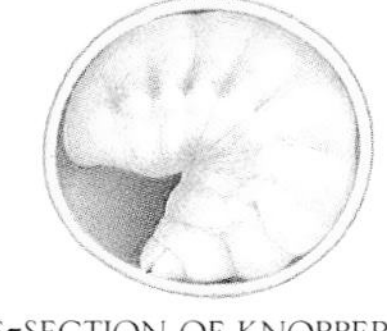

CROSS-SECTION OF KNOPPER GALL SHOWING LARVA (SHOWN X 6)

KNOPPER GALL (SHOWN X 1¼)

*All galls shown life sized unless otherwise specified*

# GALLS 2

THE GALL WASPS RESPONSIBLE for most of the galls that develop on the oak are minute insects, often only about 3 mm (⅛ in) long. They belong to the Hymenoptera order of insects, which also includes bees, other wasps, and ants. The females of the successive sexual and asexual generations cause different galls, often on different parts of the oak or even on different species of oak. Galls provide the insect larvae with an assured supply of food. The larvae are surrounded by the nutritious tissue of the host plant, and are also protected from their natural enemies and from the extremes of climate. However, numerous other species have evolved to take advantage of the same benefits. Some galls support a complex community of parasites, predators, and "cuckoo wasps", or inquilines. More than 20 different species have been recorded from oak apples.

RED-PEA GALLS

**RED-PEA GALL**
Between 10 and 20 of these round, glossy galls can be found on the underside of a single oak leaf. They are caused by the asexual generation of the gall wasp *Cynips divisa*. The adults emerge in autumn and, after overwintering, lay their eggs in leaf buds, producing small, oval "wart galls". The males and females of the sexual generation emerge in summer.

*Leaf folded to underside*

*Leaf folded to upperside*

*Macrodiplosis dryobia* EFFECT ON LEAF

*Macrodiplosis volvens* EFFECT ON LEAF

**GALL MIDGES**
These tiny flies belong to the family Cecidomyiidae. They are responsible for many familiar galls on a wide variety of plants. *Macrodiplosis* causes a crescent-shaped area of the oak leaf margin to fold, which forms the common late-summer gall.

**CHERRY GALL**
In late summer and autumn, cherry galls are conspicuous on the undersides of oak leaves. They are caused by the asexual generation of the gall wasp *Cynips quercusfolii*. After leaf fall, the asexual females remain in the galls until late winter, when they emerge to lay their eggs in dormant leaf buds. The resulting dark violet galls can be found by careful searching in the spring. The sexual generation of wasps emerges in summer.

CHERRY GALLS (SHOWN X 1¼)

GALLS OF SEXUAL GENERATION OF *Cynips quercusfolii*

*Dark violet galls are inconspicuous*

STRIPED GALLS

**STRIPED GALL**
The asexual generation of the gall wasp *Andricus longiventris* produces the striped gall. The females emerge from the leaf galls in mid-winter to lay their unfertilized eggs in dormant buds, giving rise to an inconspicuous, greenish "bud gall".

**SILK-BUTTON GALL**

Very large numbers of the silk-button gall, caused by the asexual generation of the gall wasp *Neuroterus numismalis*, can occur on the underside of a single leaf, usually towards the tip. The sexual generation causes "blister galls" on the leaf in late spring.

SILK-BUTTON GALLS

SILK-BUTTON GALL (SHOWN X 3½)

OYSTER GALLS

OYSTER GALL (SHOWN X 5)

**SPANGLE GALLS**

Four types of spangle gall can be found on the undersides of oak leaves in late summer and autumn. They are caused by the asexual generation of *Neuroterus* gall wasps. The smooth spangle gall, caused by *Neuroterus albipes*, usually occurs towards the base of the leaf, and the cupped spangle gall, caused by *Neuroterus tricolor*, can often be found on leaves of the lammas shoots (*see page 15*).

SPANGLE GALLS

CUPPED SPANGLE GALL (SHOWN X 3)

SMOOTH SPANGLE GALL (SHOWN X 3)

COMMON SPANGLE GALLS

COMMON SPANGLE GALL (SHOWN X 4)

CURRANT GALLS

**OYSTER GALL**

Oyster galls are caused by the asexual generation of the gall wasp *Andricus anthracina*. They occur in late summer on the undersides of leaves, usually attached to the midrib or a vein. The adults emerge after the gall has fallen to the ground in the autumn, and lay their eggs in dormant buds. The "bud galls" of the sexual generation appear in spring.

**COMMON SPANGLE GALL**

Up to 100 common spangle galls, caused by the gall wasp *Neuroterus quercusbaccarum*, can be found scattered all over the undersurface of a single leaf. Each spangle gall contains a single larva of the asexual generation, and falls off before leaf fall, but the larvae continue to develop over winter. The females emerge in spring, and lay their eggs in young male catkins. When mature, the resulting "currant galls" of the sexual generation resemble bunches of redcurrants. The adults emerge in summer, and the females lay their eggs on the undersides of leaves.

**OAK APPLE**

This gall is probably the most famous of all plant galls, and even has its own festival in Great Britain – Oak Apple Day on 29 May commemorates the restoration of the monarchy, and Charles II's return to England in May 1660. As its name suggests, the oak apple resembles a small apple. It is caused by the gall wasp *Biorhiza pallida*. Each gall contains up to 30 larvae of the sexual generation. The males and females emerge from separate galls in summer. After mating, the fertile females crawl into the soil, where they search out the small oak rootlets into which they insert their eggs. The resulting spherical root galls occur in clusters. Each gall contains a single larva of the asexual generation. The females are wingless, and emerge at the end of the second winter. They crawl up the trunk to lay their eggs in the young buds. More than 20 different species of inquiline, parasite, and predator have been found in oak apples.

TORTRIX MOTH *Pammenea inquilina*, AN INQUILINE OF OAK APPLES

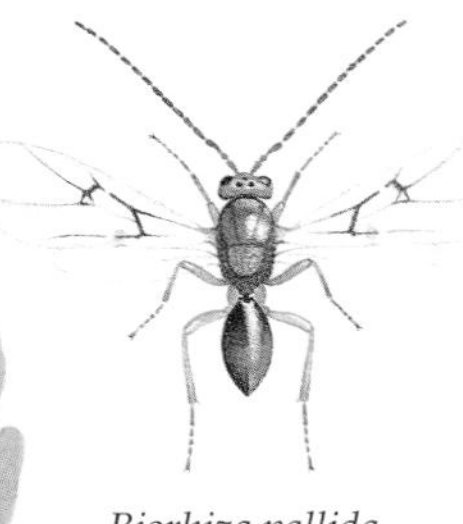

*Biorhiza pallida* (SHOWN X 4)

FEMALE OF ASEXUAL GENERATION (SHOWN X 4)

ROOT GALLS OF ASEXUAL GENERATION (SHOWN X 2)

OAK APPLE

# BUGS

THE TRUE BUGS BELONG to the large order of insects, the Hemiptera, of which about 45 occur on oak. Bugs feed by sucking the sap or juices of plants or animals with long, needle-like or short, dagger-like mouthparts. In this way they can be distinguished from beetles, which have biting mouthparts. The bugs are divided into two sub-orders, the Homoptera and the Heteroptera. All bugs have two pairs of wings (except for the females of some aphids, which are wingless). In the Homoptera the forewings are of a uniform consistency, either transparent throughout, as in the aphids, or thickened, as in the froghoppers. The forewings, or wingcases, of the Heteroptera are thickened, but with a transparent, membranous apex, which appears as a diamond-shaped tip when the wings are folded. Many bugs are small, inconspicuous insects, like the plant-lice, but others are large, colourful, and handsome animals, like the Hawthorn Bug. The aphids are among the most numerous of all insects, and several are destructive pests of agriculture.

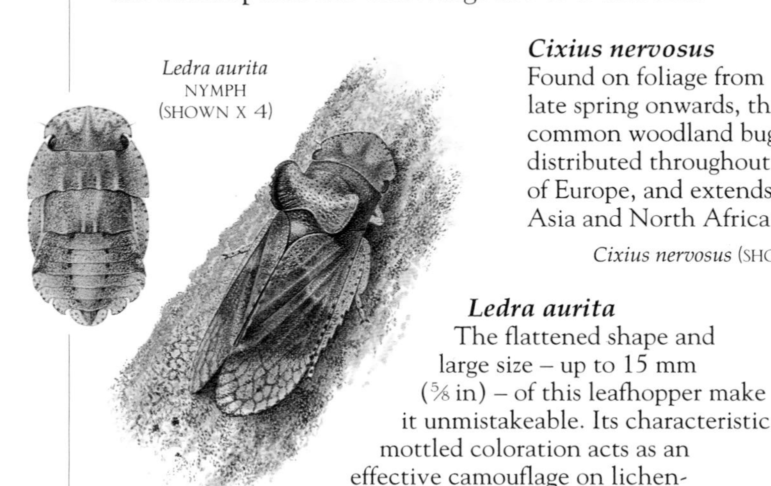

*Ledra aurita* NYMPH (SHOWN X 4)

*Ledra aurita* (SHOWN X 2½)

**Cixius nervosus**
Found on foliage from late spring onwards, this common woodland bug is distributed throughout most of Europe, and extends into Asia and North Africa.

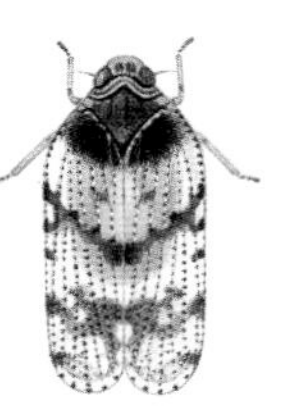

*Cixius nervosus* (SHOWN X 4)

**Ledra aurita**
The flattened shape and large size – up to 15 mm (⅝ in) – of this leafhopper make it unmistakeable. Its characteristic mottled coloration acts as an effective camouflage on lichen-covered tree branches.

**Aphrophora alni**
This common member of the froghopper family, the Cercopidae, is found on a wide variety of trees throughout most of Europe. The larva of a related species is responsible for "cuckoo spit".

*Aphrophora alni* (SHOWN X 4)

**Anthocoris confusus**
Sometimes known as a flower bug, *Anthocoris confusus* is, in fact, predatory, and feeds on mites, aphids, and other small insects. It is common on oak and other deciduous trees, and the adults can be found on these trees from mid- to late summer.

*Anthocoris confusus* (SHOWN X 7¼)

CAPSID BUGS

The capsid bugs are the most numerous family of bugs. For the most part they are small, delicate, plant-feeding insects, although several are at least partially predacious. Capsid bugs can be recognized by the pair of looped veins in the membrane at the tip of each forewing.

MALE

FEMALE

*Harpocera thoracica* (SHOWN X 4)

*Deraeocoris lutescens* (SHOWN X 6)

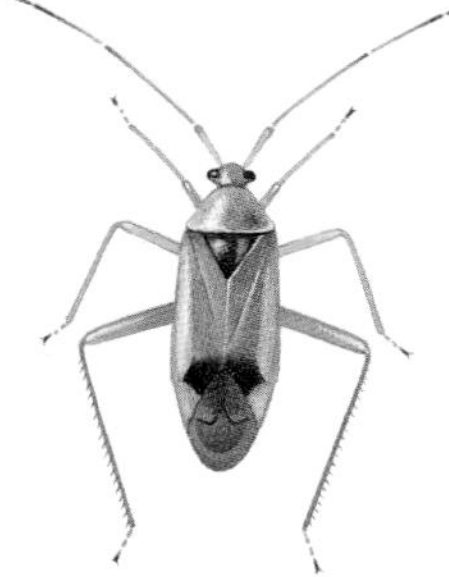

*Megacoelum infusum* (SHOWN X 3)

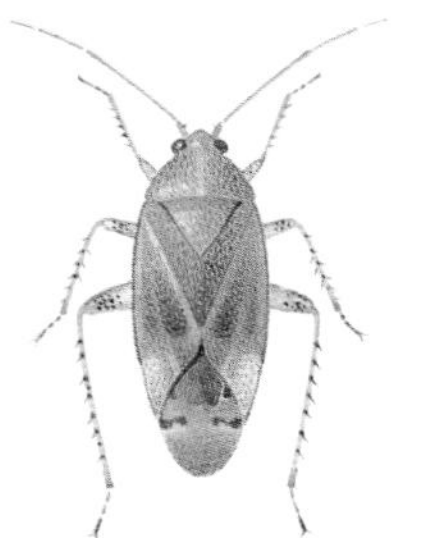

*Psallus varians* (SHOWN X 6)

*Psallus perrisi* (SHOWN X 6)

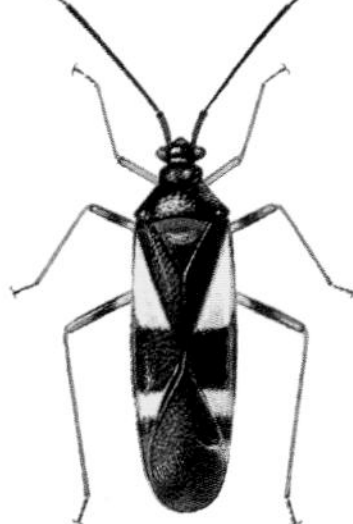

*Dryophilocoris flavoquadrimaculatus* (SHOWN X 4)

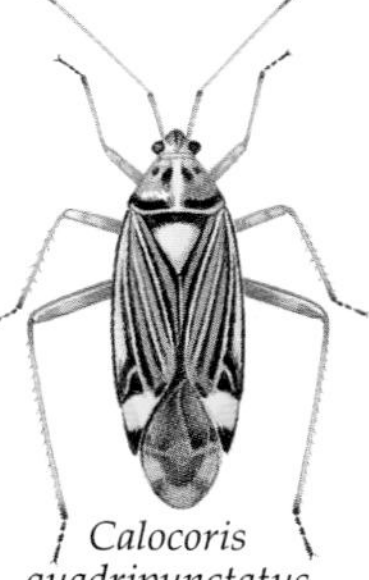

*Calocoris quadripunctatus* (SHOWN X 2½)

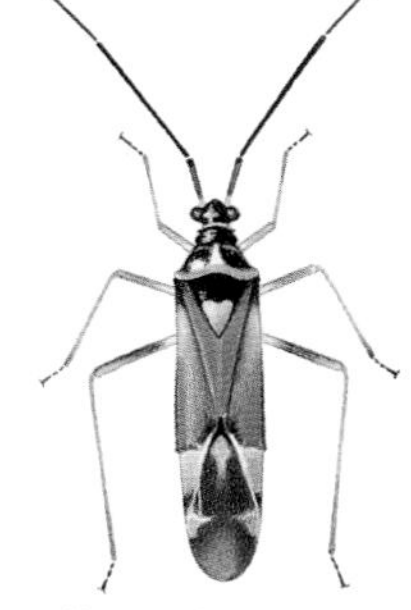

*Cyllecoris histrionicus* (SHOWN X 4)

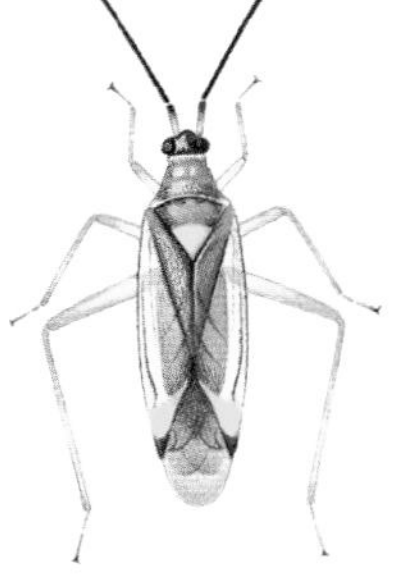

*Campyloneura virgula* (SHOWN X 6)

*Phytocoris tiliae* (SHOWN X 4)

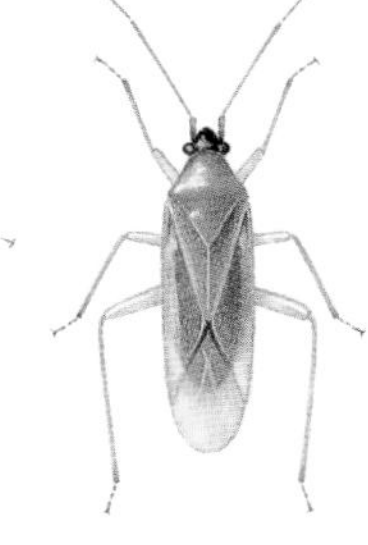

*Phylus melanocephalus* (SHOWN X 4)

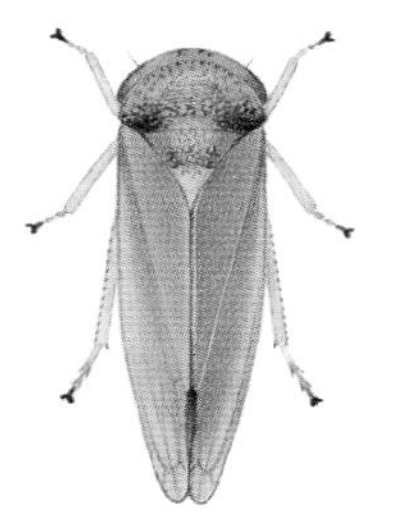

*Iassus lanio* (SHOWN X 4)

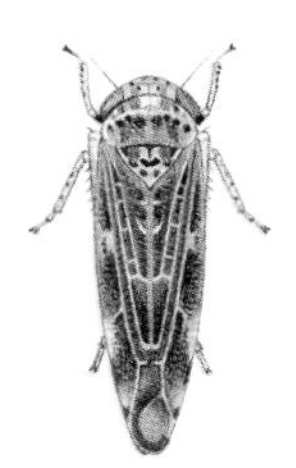

*Allygus mixtus* (SHOWN X 4)

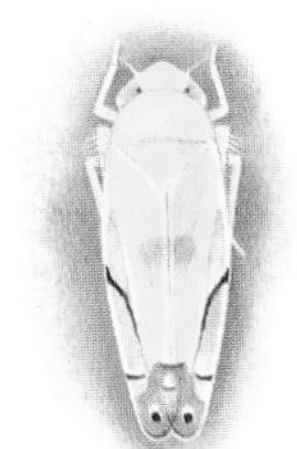

*Eurhadina pulchella* (SHOWN X 6)

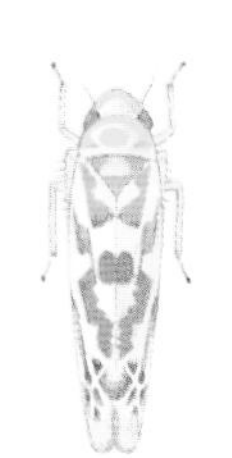

*Typhlocyba quercus* (SHOWN X 7)

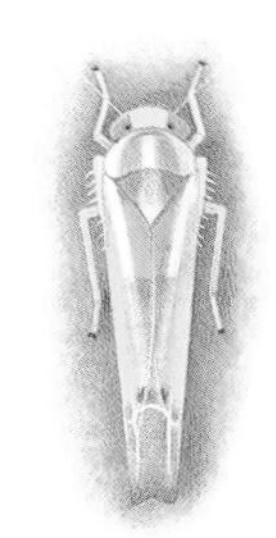

*Alebra albostriella* (SHOWN X 6)

### LEAFHOPPERS

Leafhoppers, or Cicadellidae, are small, plant-feeding insects. Many are found on trees and shrubs, and about ten species live on oak. *Iassus lanio* is a particularly distinctive, beautifully camouflaged insect. *Typhlocyba quercus* is one of the most characteristic oak tree insects. Leafhoppers can be distinguished from froghoppers by the row of spines down the outer side of their hindlegs.

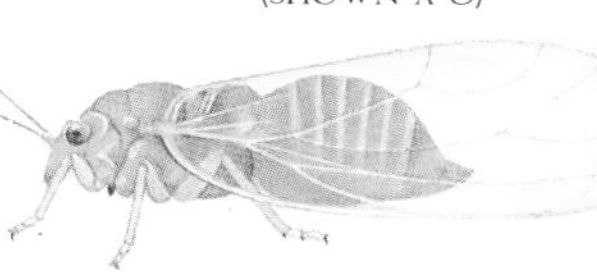

*Trioza remota* (SHOWN X 8)

### *Trioza remota*

This is the only member of the plantlouse family, or Psyllidae, specifically associated with oak. Plantlice bear a superficial resemblance to leafhoppers, but their wings are transparent, and their antennae consist of many small joints.

### *Pealius quercus*

Whiteflies such as *Pealius quercus* are tiny insects related to the plantlice. They derive their name from the dusting of white, powdery wax covering their wings. *P. quercus* is the only species of whitefly to occur on oak, and both adults and larvae feed on the undersides of leaves.

*Pealius quercus* (SHOWN X 5)

### *Empicoris vagabundus*

Active carnivores, assassin bugs such as *Empicoris vagabundus* hunt for aphids, barklice, and other small, soft-bodied insects on oak and other deciduous trees. The adults appear in mid-summer.

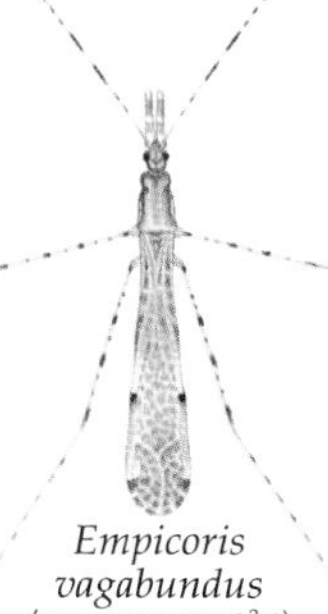

*Empicoris vagabundus* (SHOWN X 4¾)

GREEN SHIELD BUG *Palomena prasina* (SHOWN X 2½)

*Triangular, shieldlike plate covers front of abdomen*

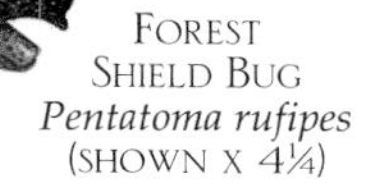

FOREST SHIELD BUG *Pentatoma rufipes* (SHOWN X 4¼)

*Troilus luridus* (SHOWN X 2½)

HAWTHORN SHIELD BUG *Acanthosoma haemorrhoidale* (SHOWN X 2½)

### SHIELD BUGS

The shield bugs of the family Pentatomidae are the largest and most handsome of all the bugs. They derive their name from their scutellum – a large, triangular, shieldlike plate that covers the front of the abdomen, and reaches the tip of the folded wing membranes. Adult shield bugs appear in mid-summer.

### APHIDS

Aphids are among the most numerous of insects. All aphids suck the sap of plants, and consequently many are serious agricultural pests. About ten different species can be found on oak, of which *Lachnus roboris* is the most distinctive.

*Thelaxes dryophila,* WINGLESS FEMALE APHID (SHOWN X 8)

*Lachnus roboris* (SHOWN X 6)

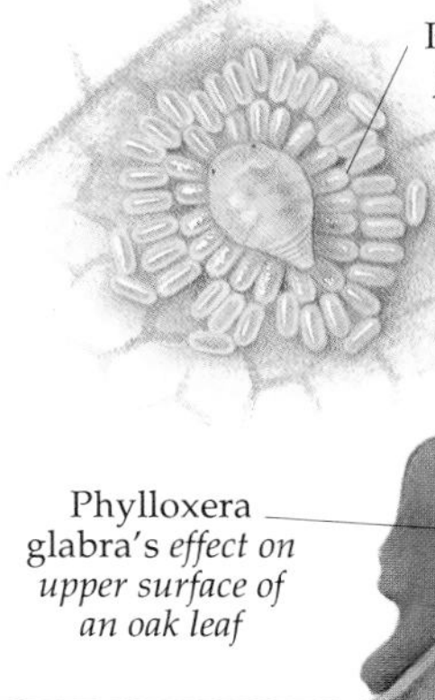

Phylloxera quercus *female with eggs on underside of leaf*

Phylloxera glabra's *effect on upper surface of an oak leaf*

# SPIDERS

IN GREEK LEGEND, THE PRINCESS Arachne was turned into a spider, and consequently, spiders belong to the class of invertebrates known as the Arachnida. The arachnids differ from insects by having four pairs of legs, and neither wings nor antennae. In the true spiders, the body is divided into two parts – the cephalothorax, which unites the head and central region of the body, and the abdomen. In the animal kingdom, only humans and spiders have the ability to construct traps separate from their bodies in order to catch prey. However, spiders' prey is not always entrapped in a web. Wolf spiders hunt, jumping spiders pounce, and crab spiders lie in ambush. Many spiders have evolved an elaborate courtship ritual, and maternal care is highly developed in some species. Oakwoods provide excellent living conditions for many of Britain's 600 species of spider.

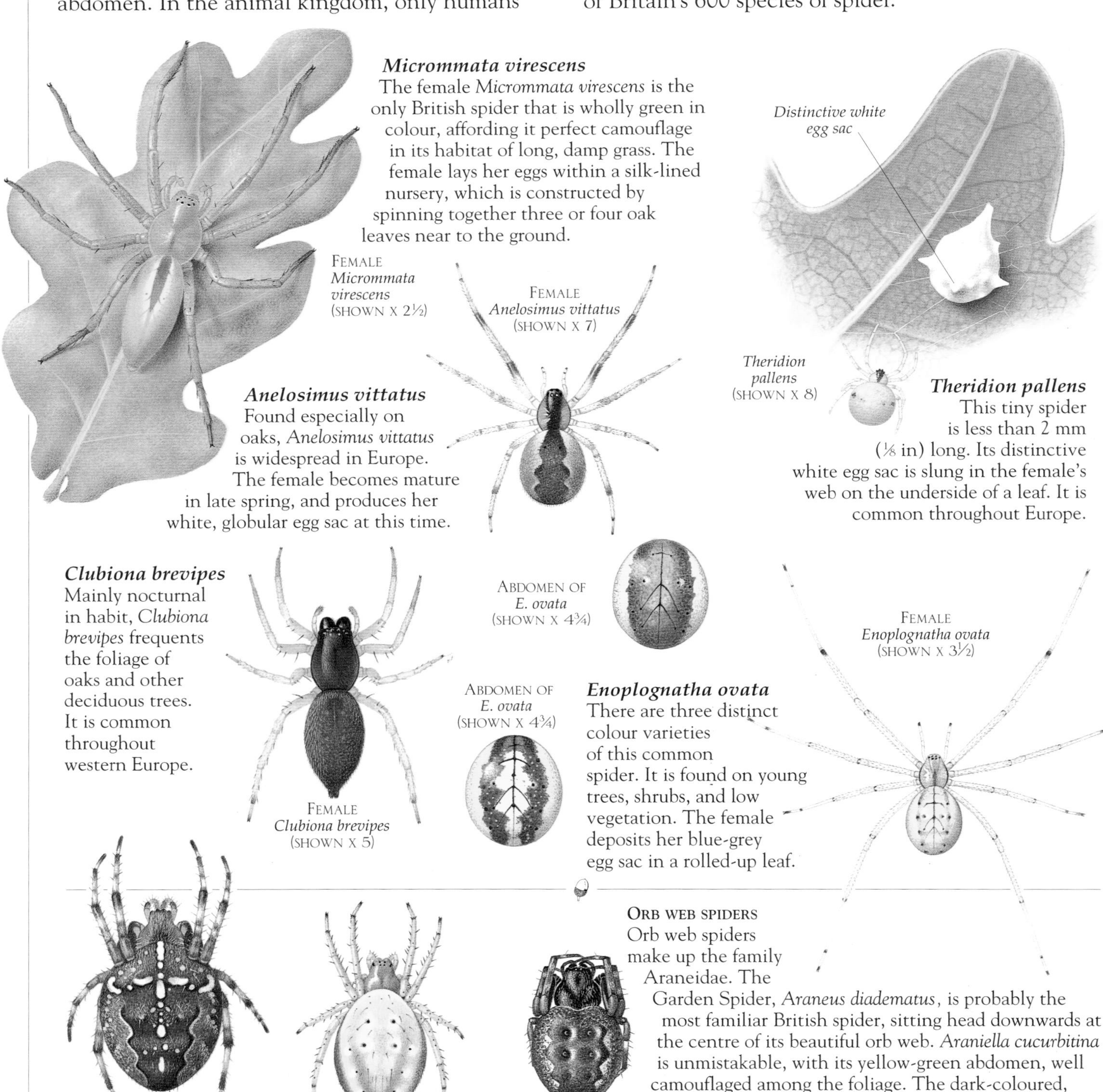


***Micrommata virescens***
The female *Micrommata virescens* is the only British spider that is wholly green in colour, affording it perfect camouflage in its habitat of long, damp grass. The female lays her eggs within a silk-lined nursery, which is constructed by spinning together three or four oak leaves near to the ground.

***Anelosimus vittatus***
Found especially on oaks, *Anelosimus vittatus* is widespread in Europe. The female becomes mature in late spring, and produces her white, globular egg sac at this time.

***Theridion pallens***
This tiny spider is less than 2 mm (⅛ in) long. Its distinctive white egg sac is slung in the female's web on the underside of a leaf. It is common throughout Europe.

***Clubiona brevipes***
Mainly nocturnal in habit, *Clubiona brevipes* frequents the foliage of oaks and other deciduous trees. It is common throughout western Europe.

***Enoplognatha ovata***
There are three distinct colour varieties of this common spider. It is found on young trees, shrubs, and low vegetation. The female deposits her blue-grey egg sac in a rolled-up leaf.

**ORB WEB SPIDERS**
Orb web spiders make up the family Araneidae. The Garden Spider, *Araneus diadematus*, is probably the most familiar British spider, sitting head downwards at the centre of its beautiful orb web. *Araniella cucurbitina* is unmistakable, with its yellow-green abdomen, well camouflaged among the foliage. The dark-coloured, flattened *Nuctenea umbratica*, on the other hand, lives concealed beneath the bark of dead trees.

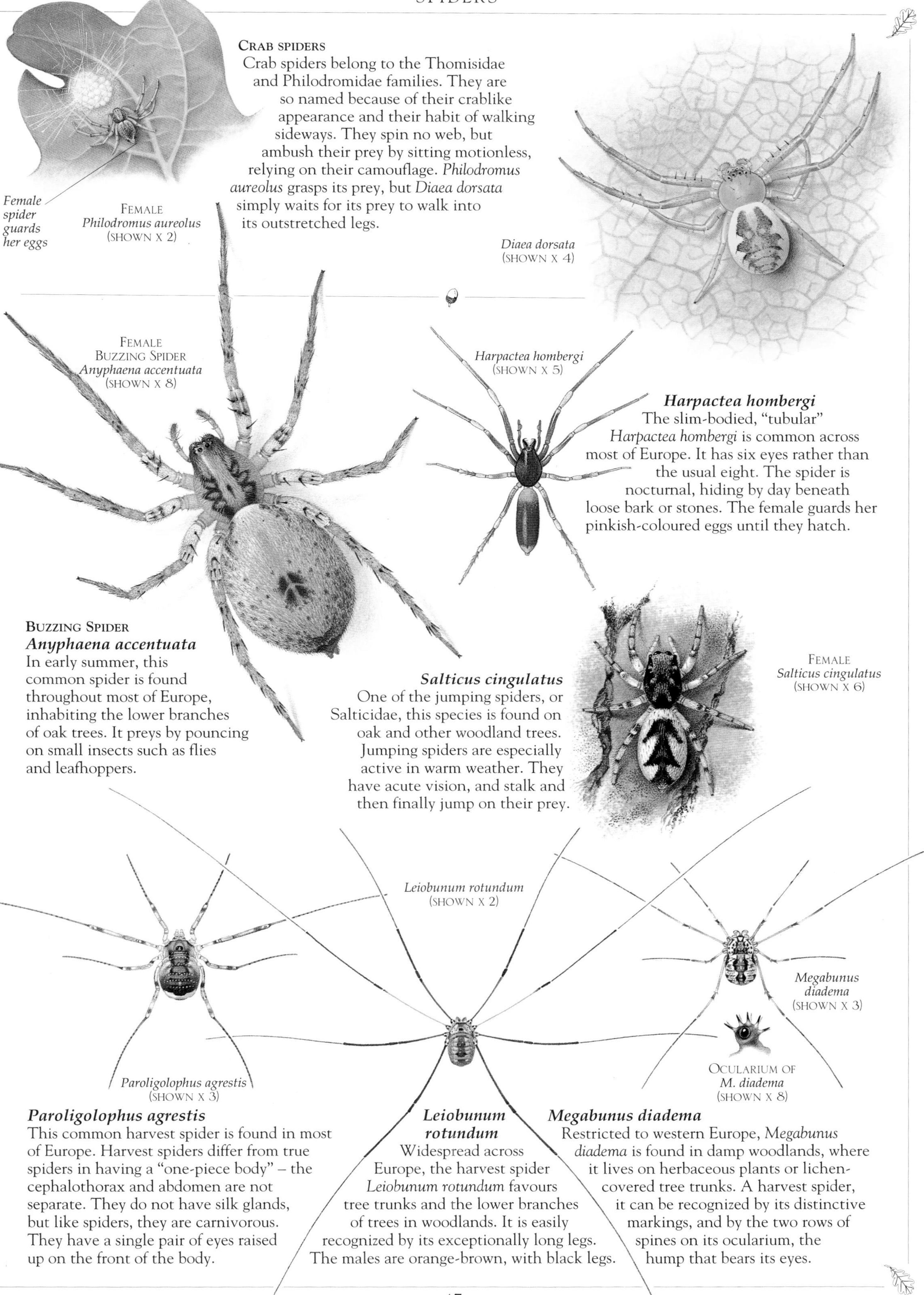

Female spider guards her eggs

FEMALE
*Philodromus aureolus*
(SHOWN X 2)

**CRAB SPIDERS**

Crab spiders belong to the Thomisidae and Philodromidae families. They are so named because of their crablike appearance and their habit of walking sideways. They spin no web, but ambush their prey by sitting motionless, relying on their camouflage. *Philodromus aureolus* grasps its prey, but *Diaea dorsata* simply waits for its prey to walk into its outstretched legs.

*Diaea dorsata*
(SHOWN X 4)

FEMALE
BUZZING SPIDER
*Anyphaena accentuata*
(SHOWN X 8)

*Harpactea hombergi*
(SHOWN X 5)

### *Harpactea hombergi*

The slim-bodied, "tubular" *Harpactea hombergi* is common across most of Europe. It has six eyes rather than the usual eight. The spider is nocturnal, hiding by day beneath loose bark or stones. The female guards her pinkish-coloured eggs until they hatch.

**BUZZING SPIDER**

### *Anyphaena accentuata*

In early summer, this common spider is found throughout most of Europe, inhabiting the lower branches of oak trees. It preys by pouncing on small insects such as flies and leafhoppers.

### *Salticus cingulatus*

One of the jumping spiders, or Salticidae, this species is found on oak and other woodland trees. Jumping spiders are especially active in warm weather. They have acute vision, and stalk and then finally jump on their prey.

FEMALE
*Salticus cingulatus*
(SHOWN X 6)

*Leiobunum rotundum*
(SHOWN X 2)

*Megabunus diadema*
(SHOWN X 3)

OCULARIUM OF
*M. diadema*
(SHOWN X 8)

*Paroligolophus agrestis*
(SHOWN X 3)

### *Paroligolophus agrestis*

This common harvest spider is found in most of Europe. Harvest spiders differ from true spiders in having a "one-piece body" – the cephalothorax and abdomen are not separate. They do not have silk glands, but like spiders, they are carnivorous. They have a single pair of eyes raised up on the front of the body.

### *Leiobunum rotundum*

Widespread across Europe, the harvest spider *Leiobunum rotundum* favours tree trunks and the lower branches of trees in woodlands. It is easily recognized by its exceptionally long legs. The males are orange-brown, with black legs.

### *Megabunus diadema*

Restricted to western Europe, *Megabunus diadema* is found in damp woodlands, where it lives on herbaceous plants or lichen-covered tree trunks. A harvest spider, it can be recognized by its distinctive markings, and by the two rows of spines on its ocularium, the hump that bears its eyes.

*All invertebrates shown life sized unless otherwise specified*

# OTHER INVERTEBRATES

OAKWOODS SUPPORT a greater diversity of invertebrates than any other habitat in Europe. Not only are there more species associated with oak trees than with any other plant, but, in addition, large numbers are dependent upon dead and decaying wood. Peel the loose bark off the trunk of an old, fallen log, and a bewildering assortment of snails, woodlice, spiders, centipedes, millipedes, beetles, springtails, and various larvae is revealed. The sheer physical complexity of a single tree provides innumerable "niches" for different species to exploit. Many are only found on venerable oaks in ancient woodland, like medieval hunting forests. Ecologists use these old woodland species to help them to identify those woods that are the most valuable for conservation.

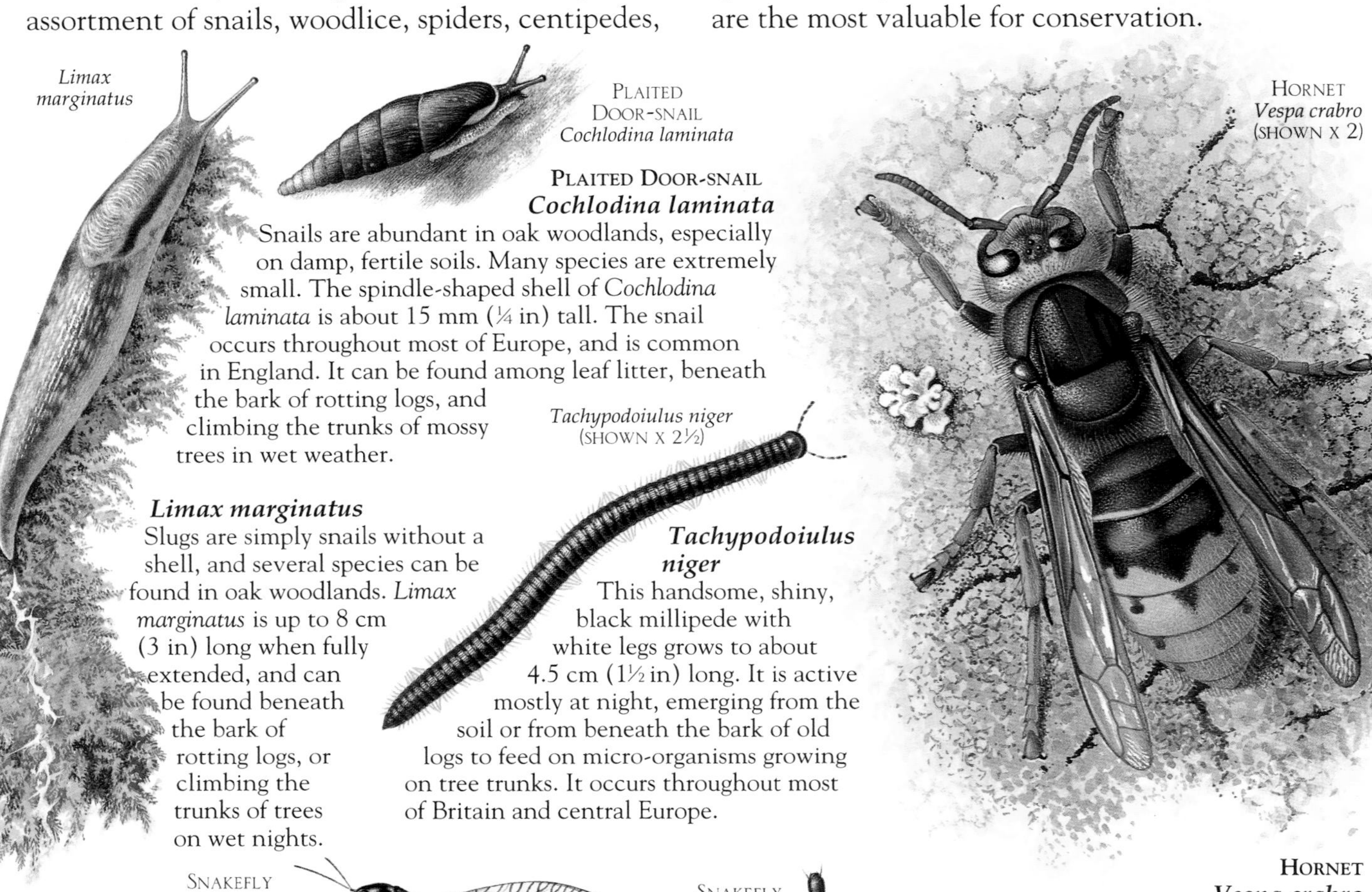


**PLAITED DOOR-SNAIL**
***Cochlodina laminata***

Snails are abundant in oak woodlands, especially on damp, fertile soils. Many species are extremely small. The spindle-shaped shell of *Cochlodina laminata* is about 15 mm (¼ in) tall. The snail occurs throughout most of Europe, and is common in England. It can be found among leaf litter, beneath the bark of rotting logs, and climbing the trunks of mossy trees in wet weather.

***Limax marginatus***

Slugs are simply snails without a shell, and several species can be found in oak woodlands. *Limax marginatus* is up to 8 cm (3 in) long when fully extended, and can be found beneath the bark of rotting logs, or climbing the trunks of trees on wet nights.

***Tachypodoiulus niger***

This handsome, shiny, black millipede with white legs grows to about 4.5 cm (1½ in) long. It is active mostly at night, emerging from the soil or from beneath the bark of old logs to feed on micro-organisms growing on tree trunks. It occurs throughout most of Britain and central Europe.

**HORNET**
***Vespa crabro***

Widespread in mainland Europe, in Britain the Hornet is confined mainly to parts of southern England. It is distinguished from common wasps by its larger size, darker colour, and brown rather than black markings. A Hornet's large, dome-shaped nest is built in a hollow tree. At the end of summer, the young queens hibernate in a secure place while the rest of the colony dies.

**SNAKEFLY**
***Raphidia notata***

The Snakefly derives its name from its arched, elongated head and thorax, which give it the appearance of a serpent ready to strike. It lives in the upper branches of oaks, where it feeds on aphids and other small insects. The larvae live in decaying wood.

**LACEWINGS**

Lacewings are delicate insects that live among trees' upper branches. They are most active during the evening, although Green Lacewings will flit about if disturbed during the day. Both the adults and the larvae feed mainly on aphids. The larger species have a ladderlike arrangment of small veins along the front edges of their wings.

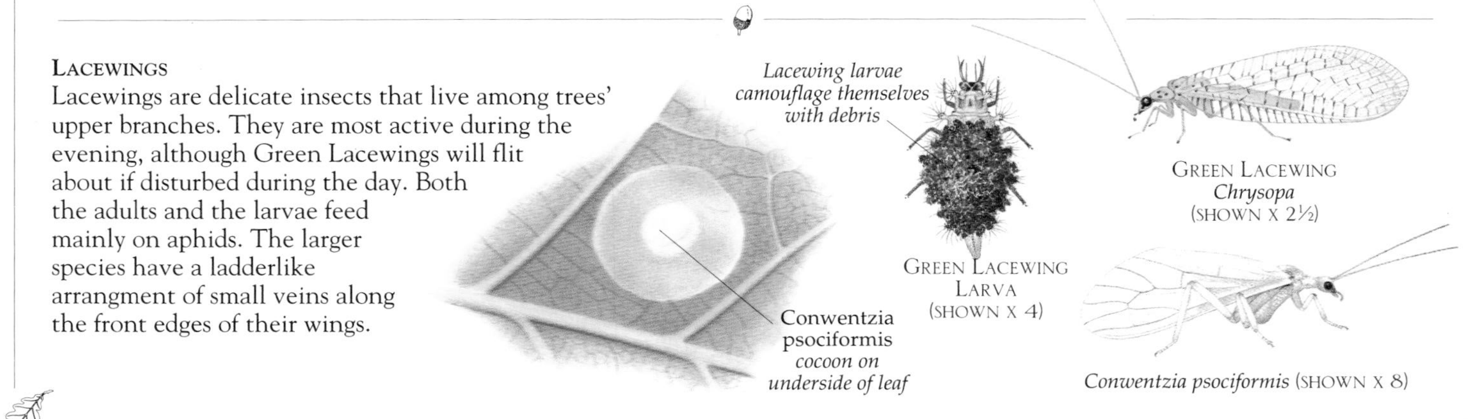

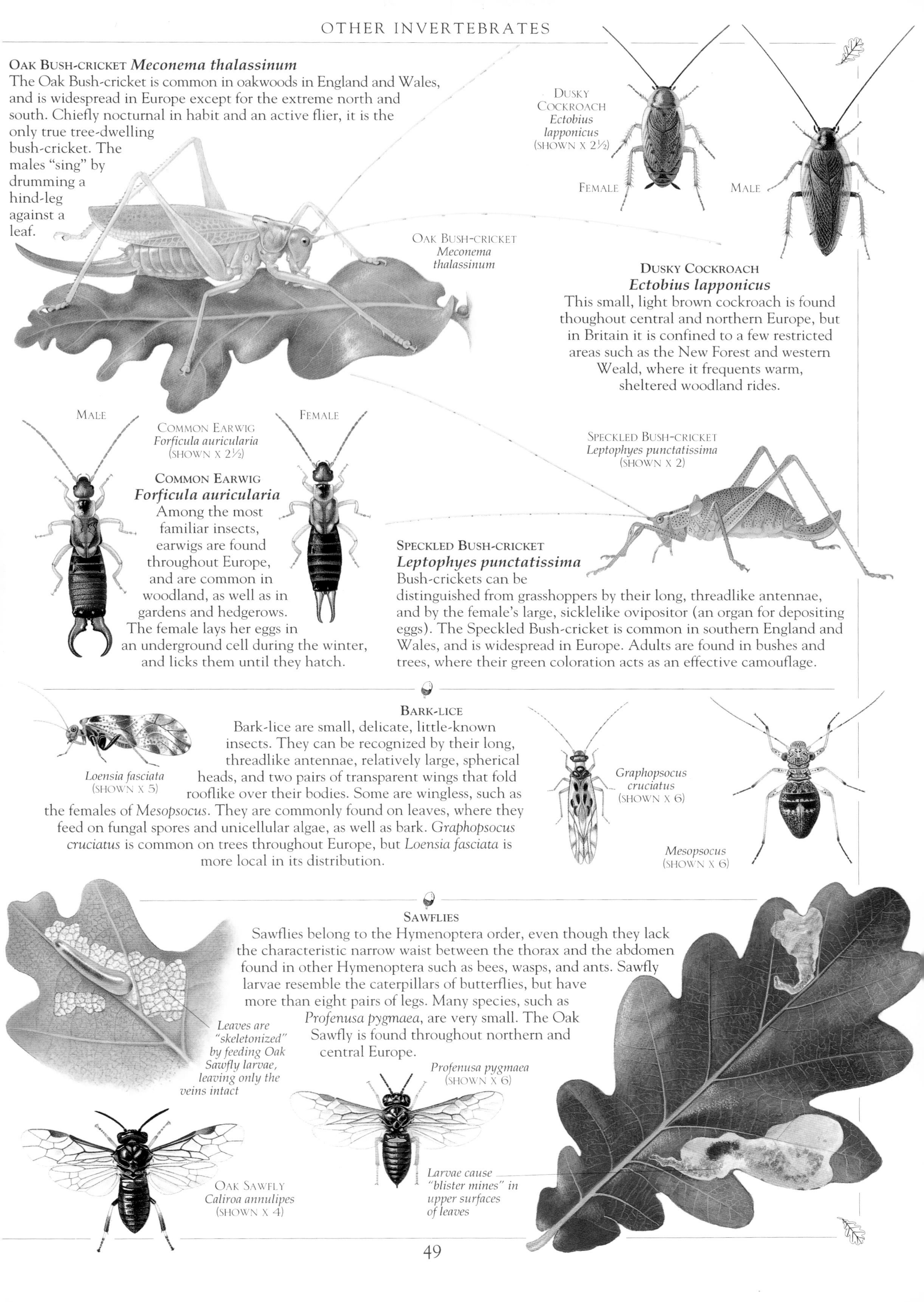

**Oak Bush-cricket *Meconema thalassinum***

The Oak Bush-cricket is common in oakwoods in England and Wales, and is widespread in Europe except for the extreme north and south. Chiefly nocturnal in habit and an active flier, it is the only true tree-dwelling bush-cricket. The males "sing" by drumming a hind-leg against a leaf.

Oak Bush-cricket *Meconema thalassinum*

Dusky Cockroach *Ectobius lapponicus* (shown x 2½)

Female — Male

**Dusky Cockroach**
***Ectobius lapponicus***

This small, light brown cockroach is found thoughout central and northern Europe, but in Britain it is confined to a few restricted areas such as the New Forest and western Weald, where it frequents warm, sheltered woodland rides.

Male — Female

Common Earwig *Forficula auricularia* (shown x 2½)

**Common Earwig**
***Forficula auricularia***

Among the most familiar insects, earwigs are found throughout Europe, and are common in woodland, as well as in gardens and hedgerows. The female lays her eggs in an underground cell during the winter, and licks them until they hatch.

Speckled Bush-cricket *Leptophyes punctatissima* (shown x 2)

**Speckled Bush-cricket**
***Leptophyes punctatissima***

Bush-crickets can be distinguished from grasshoppers by their long, threadlike antennae, and by the female's large, sicklelike ovipositor (an organ for depositing eggs). The Speckled Bush-cricket is common in southern England and Wales, and is widespread in Europe. Adults are found in bushes and trees, where their green coloration acts as an effective camouflage.

**Bark-lice**

Bark-lice are small, delicate, little-known insects. They can be recognized by their long, threadlike antennae, relatively large, spherical heads, and two pairs of transparent wings that fold rooflike over their bodies. Some are wingless, such as the females of *Mesopsocus*. They are commonly found on leaves, where they feed on fungal spores and unicellular algae, as well as bark. *Graphopsocus cruciatus* is common on trees throughout Europe, but *Loensia fasciata* is more local in its distribution.

*Loensia fasciata* (shown x 5)

*Graphopsocus cruciatus* (shown x 6)

*Mesopsocus* (shown x 6)

**Sawflies**

Sawflies belong to the Hymenoptera order, even though they lack the characteristic narrow waist between the thorax and the abdomen found in other Hymenoptera such as bees, wasps, and ants. Sawfly larvae resemble the caterpillars of butterflies, but have more than eight pairs of legs. Many species, such as *Profenusa pygmaea*, are very small. The Oak Sawfly is found throughout northern and central Europe.

*Leaves are "skeletonized" by feeding Oak Sawfly larvae, leaving only the veins intact*

Oak Sawfly *Caliroa annulipes* (shown x 4)

*Profenusa pygmaea* (shown x 6)

*Larvae cause "blister mines" in upper surfaces of leaves*

*All fungi shown life sized*

# MUSHROOMS & TOADSTOOLS

IN THE WARM, MOIST DAYS of autumn, oakwoods produce a rich crop of fungi. Neither plant nor animal, fungi are essential to oakwoods. Unlike plants, they possess no chlorophyll, and are therefore unable to use the sun's energy to photosynthesize food. Instead, they live either parasitically on living plants or animals, or on their dead remains, as saprophytes. The toadstool is the visible fruit body of the fungus. The main part of the fungus is the mycelium, a network of innumerable microscopic threads, called hyphae, that grow through the soil or a host plant. One gram (1/25 oz) of soil can contain up to 100 metres (109 yds) of hyphae. Many woodland fungi infect the roots of trees, forming a mutually beneficial association called a mycorrhiza. The fungus obtains carbohydrates from the tree, and the tree's uptake of essential nutrients is improved by the fungus.

SPINDLE SHANK
*Collybia fusipes*

RUSSET SHANK
*Collybia dryophila*

**SPINDLE SHANK**
***Collybia fusipes***
The Spindle Shank's stem, or shank, is deeply grooved and often twisted, resembling a spindle of yarn. The stems join underground into a long, blackish false root, or pseudorhiza. Clusters of Spindle Shanks appear at the bases of oaks and beeches across Britain and most of mainland Europe from spring to autumn. They are not edible.

**RUSSET SHANK**
***Collybia dryophila***
A common toadstool of mixed broadleaved and oak woodlands, the Russet Shank grows on leaf litter and soil. It probably forms mycorrhizal associations with trees. The Russet Shank appears from spring to autumn across Europe, and is not edible.

*Mycena inclinata*

***Mycena inclinata***
Species of *Mycena* are generally small toadstools with slender stems. *Mycena inclinata* fits this description, and has a finely toothed edge to its cap and a densely downy stem that is dark towards the base. It grows in tufts on the stumps or trunks of dead oaks throughout Britain and mainland Europe. Appearing in the late summer and autumn, it is not edible.

**OAK MILK CAP**
***Lactarius quietus***
This is a common autumn toadstool, and one of the fungi most characteristic of oakwoods, especially those on acid soil. *Lactarius* species are among the most important of the woodland mycorrhizal fungi. The Oak Milk Cap is not edible.

OAK MILK CAP
*Lactarius quietus*

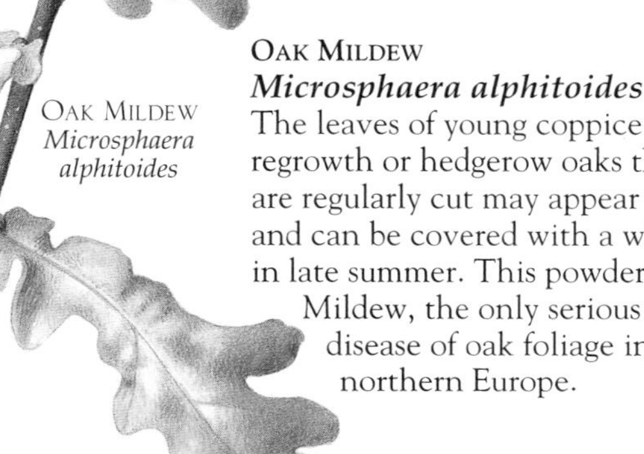

OAK MILDEW
*Microsphaera alphitoides*

**OAK MILDEW**
***Microsphaera alphitoides***
The leaves of young coppice regrowth or hedgerow oaks that are regularly cut may appear wrinkled and can be covered with a white bloom in late summer. This powder is Oak Mildew, the only serious fungal disease of oak foliage in northern Europe.

*Lactarius chrysorrheus*

***Lactarius chrysorrheus***
Several species of *Lactarius* are common in oakwoods, all closely related and similar in appearance. All of them exude a milky juice when the cap is broken, and *Lactarius chrysorrheus* can be identified by the way that its milk turns yellow when exposed to the air. It is not edible.

### Green Wood Cup

**_Chlorociboria (syn. Chlorospenium) aeruginascens_**

On woodland walks, one often comes across bright green, rotting wood. The green stain is caused by the Green Wood Cup fungus. This infects the dead wood of several broadleaved trees, but is most common on oak, hence its other name of "green oak". Wood stained by the fungus was once used in patterned veneers of various coloured woods.

### *Calocera cornea*

In late summer and autumn, dead wood often sprouts groups of yellow, gelatinous teeth about 10 mm (½ in) high. In spite of their appearance, these *Calocera* species belong to the same group of fungi as the more conventional toadstools. *Calocera cornea* tapers to a point, while the rarer *Calocera glossoides* is smaller and broader, with a rounded tip.

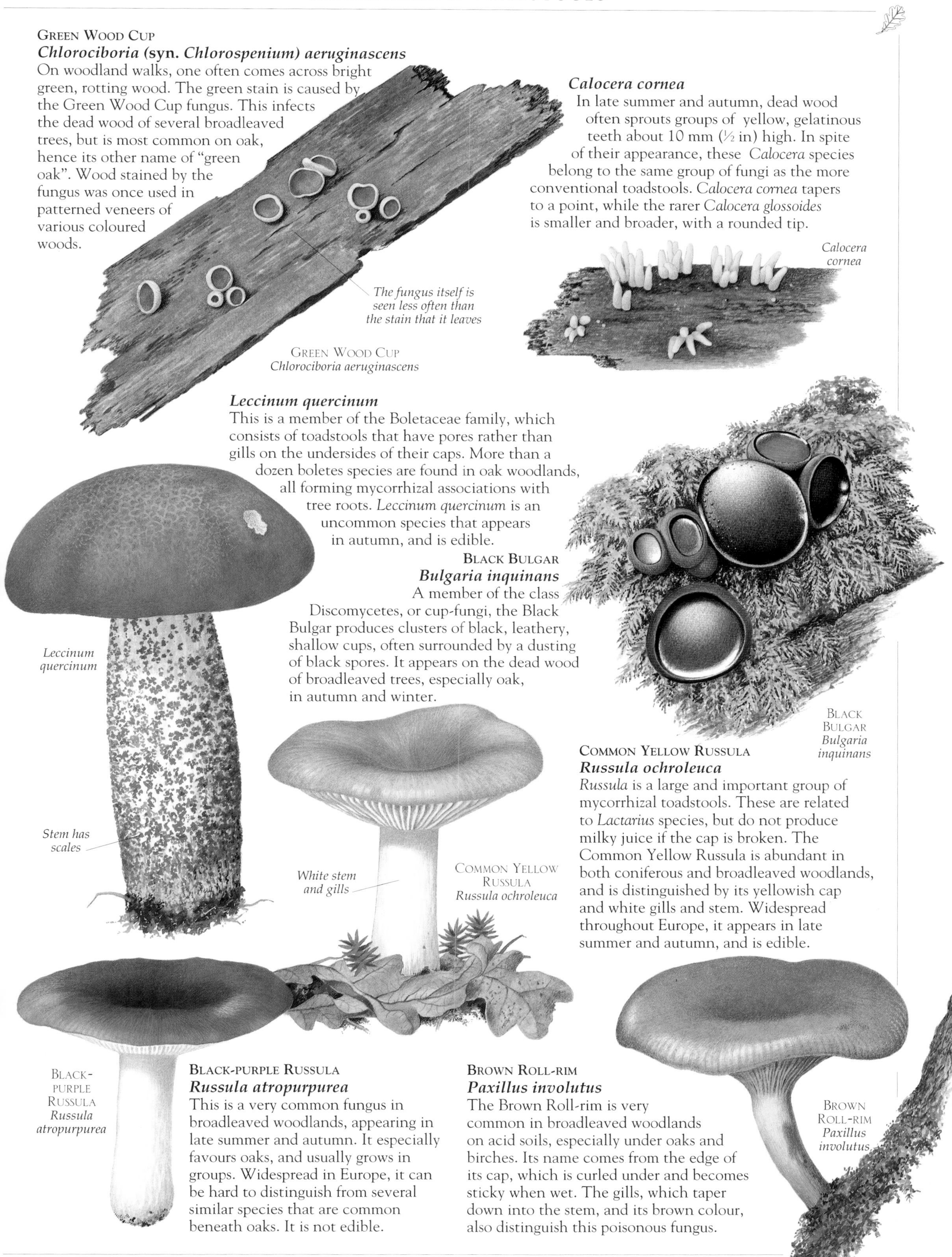


### *Leccinum quercinum*

This is a member of the Boletaceae family, which consists of toadstools that have pores rather than gills on the undersides of their caps. More than a dozen boletes species are found in oak woodlands, all forming mycorrhizal associations with tree roots. *Leccinum quercinum* is an uncommon species that appears in autumn, and is edible.

### Black Bulgar

**_Bulgaria inquinans_**

A member of the class Discomycetes, or cup-fungi, the Black Bulgar produces clusters of black, leathery, shallow cups, often surrounded by a dusting of black spores. It appears on the dead wood of broadleaved trees, especially oak, in autumn and winter.

### Common Yellow Russula

**_Russula ochroleuca_**

*Russula* is a large and important group of mycorrhizal toadstools. These are related to *Lactarius* species, but do not produce milky juice if the cap is broken. The Common Yellow Russula is abundant in both coniferous and broadleaved woodlands, and is distinguished by its yellowish cap and white gills and stem. Widespread throughout Europe, it appears in late summer and autumn, and is edible.

### Black-purple Russula

**_Russula atropurpurea_**

This is a very common fungus in broadleaved woodlands, appearing in late summer and autumn. It especially favours oaks, and usually grows in groups. Widespread in Europe, it can be hard to distinguish from several similar species that are common beneath oaks. It is not edible.

### Brown Roll-rim

**_Paxillus involutus_**

The Brown Roll-rim is very common in broadleaved woodlands on acid soils, especially under oaks and birches. Its name comes from the edge of its cap, which is curled under and becomes sticky when wet. The gills, which taper down into the stem, and its brown colour, also distinguish this poisonous fungus.

# BRACKET FUNGI

BRACKET FUNGI ARE AMONG the most spectacular of all fungi. Most have pores under their caps rather than gills, giving a spongelike appearance, which belies the fact that some of them are tough and leathery, or even hard. Many bracket fungi support their own special invertebrate fauna of beetles and fungus gnat larvae, and some even combine their bizarre appearance with a culinary reputation. Like toadstools, bracket fungi are the fruit body of the organism; the mycelium lives in the host tree. Fungi are the only organisms capable of digesting lignin, the substance of which wood is composed. They therefore play a vital part in the decomposition of dead wood, and may be found on the decaying remains of trees. A few species of bracket fungi can also damage living trees.

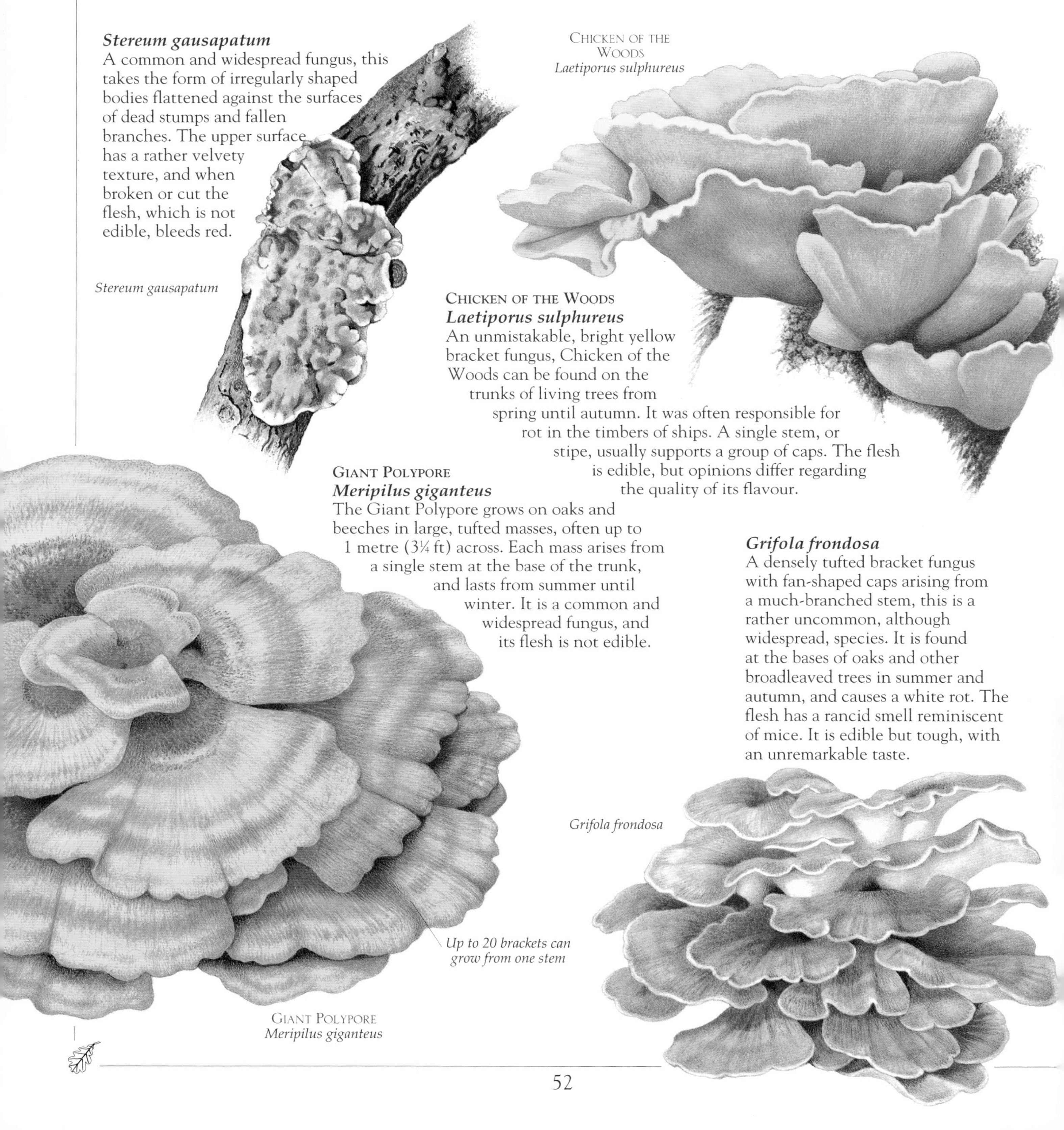

***Stereum gausapatum***
A common and widespread fungus, this takes the form of irregularly shaped bodies flattened against the surfaces of dead stumps and fallen branches. The upper surface has a rather velvety texture, and when broken or cut the flesh, which is not edible, bleeds red.

*Stereum gausapatum*

CHICKEN OF THE WOODS
*Laetiporus sulphureus*

CHICKEN OF THE WOODS
***Laetiporus sulphureus***
An unmistakable, bright yellow bracket fungus, Chicken of the Woods can be found on the trunks of living trees from spring until autumn. It was often responsible for rot in the timbers of ships. A single stem, or stipe, usually supports a group of caps. The flesh is edible, but opinions differ regarding the quality of its flavour.

GIANT POLYPORE
***Meripilus giganteus***
The Giant Polypore grows on oaks and beeches in large, tufted masses, often up to 1 metre (3¼ ft) across. Each mass arises from a single stem at the base of the trunk, and lasts from summer until winter. It is a common and widespread fungus, and its flesh is not edible.

***Grifola frondosa***
A densely tufted bracket fungus with fan-shaped caps arising from a much-branched stem, this is a rather uncommon, although widespread, species. It is found at the bases of oaks and other broadleaved trees in summer and autumn, and causes a white rot. The flesh has a rancid smell reminiscent of mice. It is edible but tough, with an unremarkable taste.

*Grifola frondosa*

*Up to 20 brackets can grow from one stem*

GIANT POLYPORE
*Meripilus giganteus*

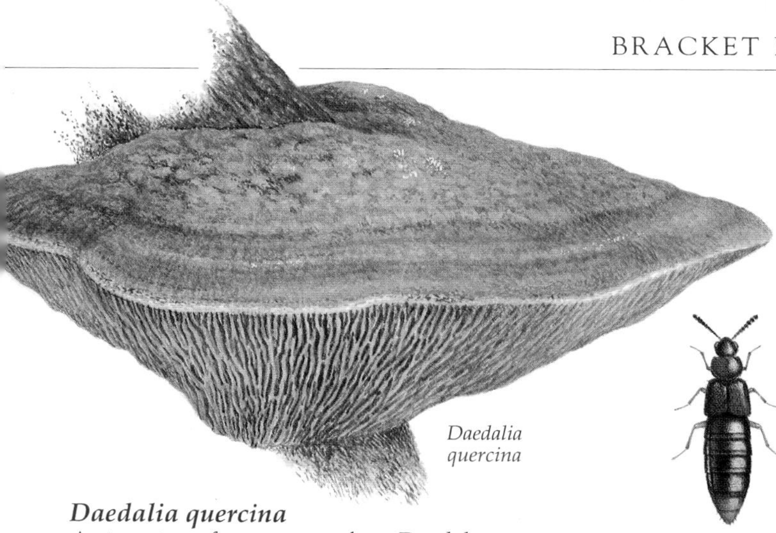

*Daedalia quercina*

### *Daedalia quercina*

As its scientific name implies, *Daedalia quercina* grows almost solely on oaks, either on stumps and cut timber, or on dead wood on living trees. Common in Britain and widespread throughout most of Europe, the inedible bracket may be found throughout the year. It is recognized by its wavy, elongated pores, which have earned it the name of Maze Gill. The Rove Beetle, *Gyrophaena stictula,* lives exclusively on this fungus.

ROVE BEETLE
*Gyrophaena stictula*
(SHOWN X 10)

### BEEFSTEAK FUNGUS *Fistulina hepatica*

This fungus owes its common name to its soft, juicy, pink flesh, which bleeds a reddish juice when cut. The bracket, which is found low down on oak or chestnut trunks from late summer, is edible when boiled, but not worth a second try.

BEEFSTEAK FUNGUS
*Fistulina hepatica*

### MANY-ZONED POLYPORE *Coriolus ( syn. Trametes) versicolor*

This distinctive, velvety, concentrically marked bracket can be found from spring to winter throughout Europe. It is common on the dead wood of deciduous trees, especially oak and beech, and is a problem in wooden fencing, causing serious decay. The flesh is not edible.

MANY-ZONED POLYPORE
*Coriolus versicolor*

### *Inonotus dryadeus*

This fungus is rather uncommon in Britain, but widespread in central and southern Europe. The bracket can be found throughout the year, low down on the trunks of oaks and other broadleaved species, and is not edible.

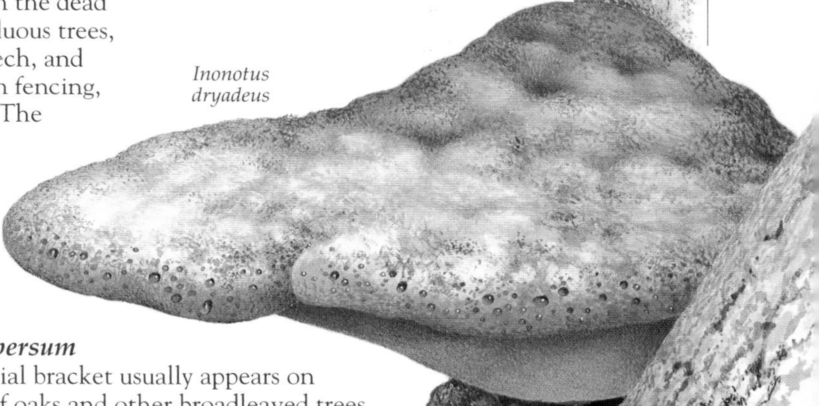

*Inonotus dryadeus*

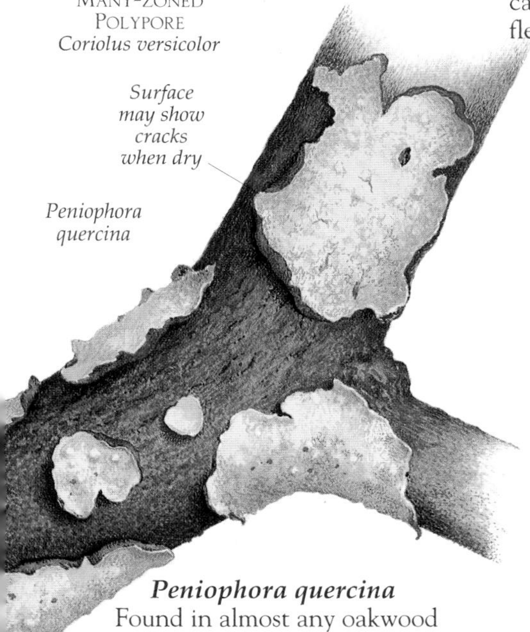

*Surface may show cracks when dry*

*Peniophora quercina*

### *Ganoderma adspersum*

This large, perennial bracket usually appears on the lower trunks of oaks and other broadleaved trees, and causes a white rot. It is widespread, and quite common on isolated trees in parks and gardens. The closely related and similar *Ganoderma applanatum* is less common, and is usually found on the trunks of old beeches. Neither species is edible.

### *Peniophora quercina*

Found in almost any oakwood at any time of year, this fungus is common on the dead and fallen branches of the oak. The irregularly lobed body is closely attached to the surface of the wood, just curling up at the edges to show a black underside, and is not edible. Many similar, related species appear on the dead wood of other trees and shrubs.

*Ganoderma adspersum*

# LICHENS

LICHENS ARE FUNGI that have evolved an intimate association with a primitive green plant, either a green alga or a cyanobacterium. Photosynthesis of the alga provides the fungus with the energy needed for growth and reproduction, while the fungus provides the alga with nutrients and protects it from extreme conditions. There are thousands of different lichens, over 300 of which have been found on oak trees. They display an astonishing diversity of form, from shrubby, branched species with a single point of attachment (fruticose) and flattened species with leaf-like lobes (foliose), to encrusting species with a thallus, or main body area, that adheres to bark or rock (crustose). Lichens are sensitive to atmospheric pollution, and several have become scarce this century.

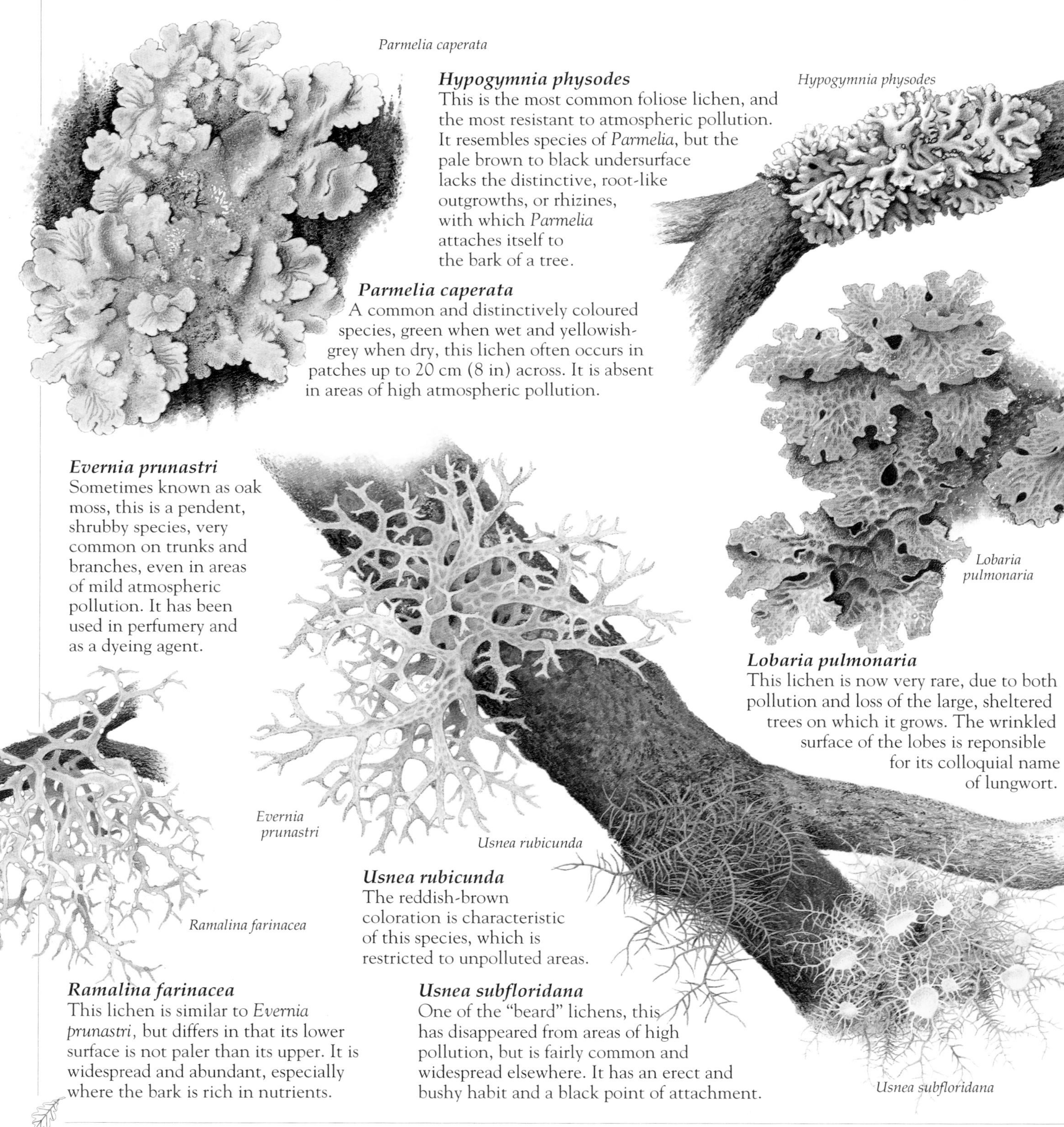


***Hypogymnia physodes***
This is the most common foliose lichen, and the most resistant to atmospheric pollution. It resembles species of *Parmelia*, but the pale brown to black undersurface lacks the distinctive, root-like outgrowths, or rhizines, with which *Parmelia* attaches itself to the bark of a tree.

***Parmelia caperata***
A common and distinctively coloured species, green when wet and yellowish-grey when dry, this lichen often occurs in patches up to 20 cm (8 in) across. It is absent in areas of high atmospheric pollution.

***Evernia prunastri***
Sometimes known as oak moss, this is a pendent, shrubby species, very common on trunks and branches, even in areas of mild atmospheric pollution. It has been used in perfumery and as a dyeing agent.

***Lobaria pulmonaria***
This lichen is now very rare, due to both pollution and loss of the large, sheltered trees on which it grows. The wrinkled surface of the lobes is reponsible for its colloquial name of lungwort.

***Usnea rubicunda***
The reddish-brown coloration is characteristic of this species, which is restricted to unpolluted areas.

***Ramalina farinacea***
This lichen is similar to *Evernia prunastri*, but differs in that its lower surface is not paler than its upper. It is widespread and abundant, especially where the bark is rich in nutrients.

***Usnea subfloridana***
One of the "beard" lichens, this has disappeared from areas of high pollution, but is fairly common and widespread elsewhere. It has an erect and bushy habit and a black point of attachment.

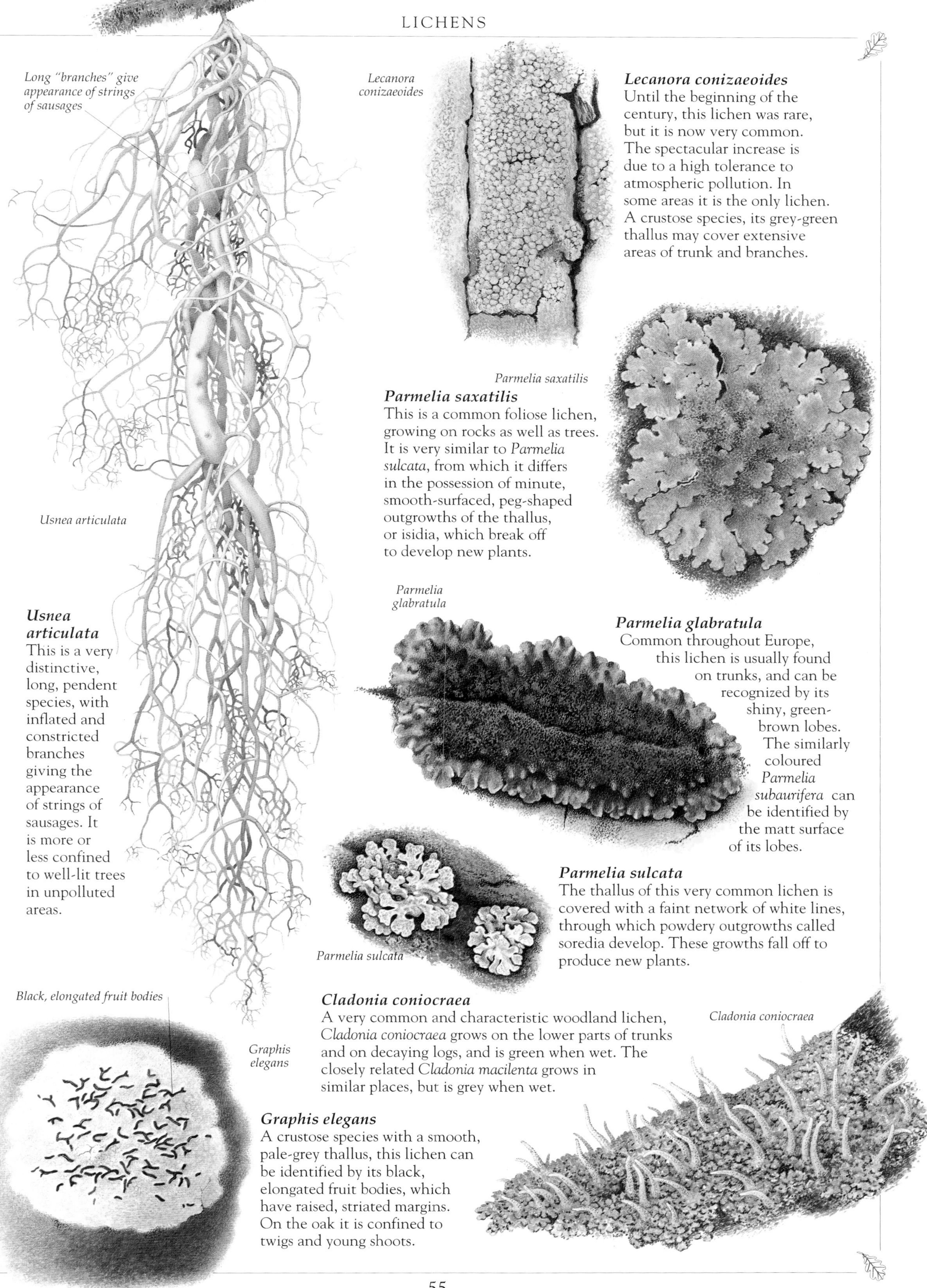


***Lecanora conizaeoides***
Until the beginning of the century, this lichen was rare, but it is now very common. The spectacular increase is due to a high tolerance to atmospheric pollution. In some areas it is the only lichen. A crustose species, its grey-green thallus may cover extensive areas of trunk and branches.

***Parmelia saxatilis***
This is a common foliose lichen, growing on rocks as well as trees. It is very similar to *Parmelia sulcata*, from which it differs in the possession of minute, smooth-surfaced, peg-shaped outgrowths of the thallus, or isidia, which break off to develop new plants.

***Usnea articulata***
This is a very distinctive, long, pendent species, with inflated and constricted branches giving the appearance of strings of sausages. It is more or less confined to well-lit trees in unpolluted areas.

***Parmelia glabratula***
Common throughout Europe, this lichen is usually found on trunks, and can be recognized by its shiny, green-brown lobes. The similarly coloured *Parmelia subaurifera* can be identified by the matt surface of its lobes.

***Parmelia sulcata***
The thallus of this very common lichen is covered with a faint network of white lines, through which powdery outgrowths called soredia develop. These growths fall off to produce new plants.

***Cladonia coniocraea***
A very common and characteristic woodland lichen, *Cladonia coniocraea* grows on the lower parts of trunks and on decaying logs, and is green when wet. The closely related *Cladonia macilenta* grows in similar places, but is grey when wet.

***Graphis elegans***
A crustose species with a smooth, pale-grey thallus, this lichen can be identified by its black, elongated fruit bodies, which have raised, striated margins. On the oak it is confined to twigs and young shoots.

# FERNS, MOSSES, & LIVERWORTS

FERNS, MOSSES, AND LIVERWORTS are flowerless plants, reproducing by small, dust-like spores that are dispersed by the wind. They are most abundant in damp habitats and moist climates, as their thin leaves are easily damaged by desiccation. Also, unlike flowering plants that produce pollen, fertilization is brought about by a male sperm, which needs a film of water in which to swim to the egg. Mosses and liverworts belong to a group of plants known as the Bryophyta. Their spores develop in a capsule, usually situated at the top of a stalk, or seta. Mosses differ from liverworts in the structure of the spore capsule and the arrangement of their leaves. About 65 different mosses and liverworts have been found growing on the bark of oaks, but very few ferns habitually grow on the trees themselves.

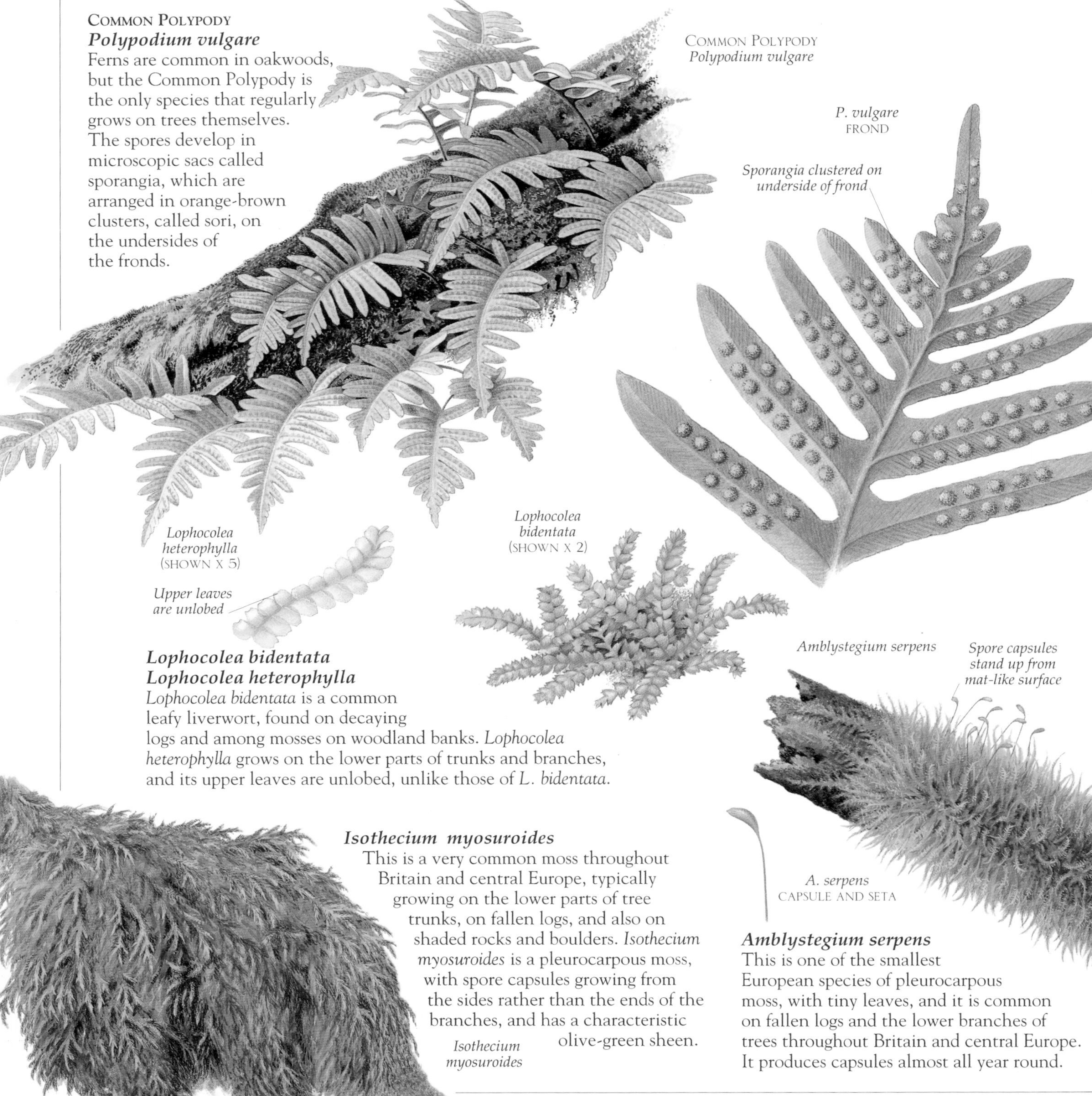


**COMMON POLYPODY**
***Polypodium vulgare***
Ferns are common in oakwoods, but the Common Polypody is the only species that regularly grows on trees themselves. The spores develop in microscopic sacs called sporangia, which are arranged in orange-brown clusters, called sori, on the undersides of the fronds.

***Lophocolea bidentata***
***Lophocolea heterophylla***
*Lophocolea bidentata* is a common leafy liverwort, found on decaying logs and among mosses on woodland banks. *Lophocolea heterophylla* grows on the lower parts of trunks and branches, and its upper leaves are unlobed, unlike those of *L. bidentata*.

***Isothecium myosuroides***
This is a very common moss throughout Britain and central Europe, typically growing on the lower parts of tree trunks, on fallen logs, and also on shaded rocks and boulders. *Isothecium myosuroides* is a pleurocarpous moss, with spore capsules growing from the sides rather than the ends of the branches, and has a characteristic olive-green sheen.

***Amblystegium serpens***
This is one of the smallest European species of pleurocarpous moss, with tiny leaves, and it is common on fallen logs and the lower branches of trees throughout Britain and central Europe. It produces capsules almost all year round.

***Dicranoweisia cirrata***
This is the most common of the acrocarpous mosses, (mosses that have capsules at the tips of erect shoots), of tree trunks and lower branches. It is usually fertile, and is found throughout Europe, especially in areas with low levels of atmospheric pollution.

*Spore capsules grow at tips of shoots*

*Dicranoweisia cirrata*

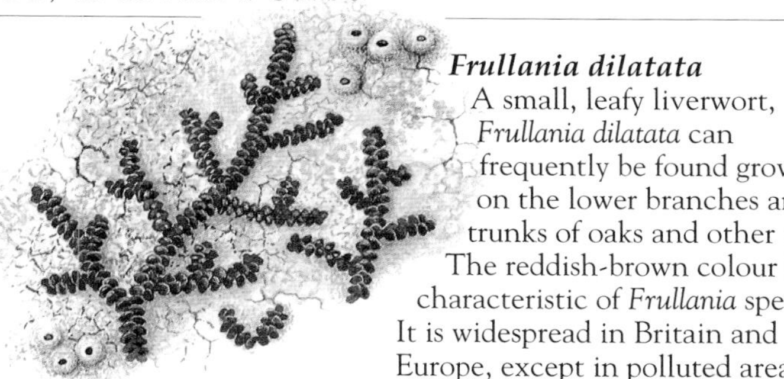

*Frullania dilatata*

***Frullania dilatata***
A small, leafy liverwort, *Frullania dilatata* can frequently be found growing on the lower branches and trunks of oaks and other trees. The reddish-brown colour is characteristic of *Frullania* species. It is widespread in Britain and Europe, except in polluted areas and the extreme north.

***Ulota crispa***
This occurs in small, yellow-green tufts on twigs and small branches of hazel and elder, as well as young oaks. Widespread throughout Europe, it is less common in the Midlands and southeast regions of Britain.

*Ulota crispa*

*U. crispa*
RIBBED CAPSULE
(SHOWN X 8)

***Dicranum scoparium***
This is a common moss on fallen logs, woodland banks, and the trunks and lower branches of trees. It grows in bright green patches and the tips of its erect shoots all curve over in the same direction.

*Dicranum scoparium*

***Hypnum cupressiforme***
A very common moss throughout Europe, this grows in mats on banks, trees, and decaying logs. Similar to *Isothecium myosuroides*, it grows higher up on trees and has more curved leaves.

*H. cupressiforme*
CAPSULE

*Sickle-shaped leaves are characteristic of species*

*Hypnum cupressiforme*

***Mnium hornum***
This is one of the most common European woodland mosses. It grows on tree trunks, decaying logs, stream banks, and woodland floors, especially on acid soil throughout Britain and central Europe. It forms extensive, dark-green patches.

TIP OF MALE SHOOT
(SHOWN X 2½)

*Mnium hornum*

***Brachythecium rutabulum***
This is a robust species of pleurocarpous moss with a characteristic yellow-green, glossy tinge to the branch tips. It is common on tree trunks and decaying logs, especially where the soil is fertile. It is abundant throughout central Europe.

*Yellow-green colour is characteristic of species*

***Eurhynchium praelongum***
A delicate species of pleurocarpous moss common in central Europe, *Eurhynchium praelongum* favours decaying stumps and logs. It has fine, feathery branches and small leaves, and regular branching of shoots is characteristic.

*Eurhynchium praelongum*

*Brachythecium rutabulum*

# THE LEAF LITTER

EACH AUTUMN, THE OAK TREES in an oakwood shed about 23 million leaves per hectare (2½ acres) – equivalent to about 2.5 tonnes per hectare (1 ton per acre). Within about nine months, this blanket of leaf litter has completely decayed to form a dark, fertile humus, returning to the soil the essential nutrients, like nitrogen, phosphorus, and potassium, taken up by the trees in the previous season. Hence the fertility of the woodland soil is maintained. This prodigious feat of natural biotechnology is brought about by the huge population of micro-organisms and invertebrates that inhabit the soil. Wood is attacked by fungi, leaves are chewed by enormous armies of mites, springtails, and woodlice, earthworms mix the products, and fungi and bacteria perform the final stages of the process. A gram (1/25 oz) of woodland soil may contain a thousand million bacteria.

*Porcellio scaber* (SHOWN X 3)

*Oniscus asellus* (SHOWN X 3)

**WOODLICE**
Woodlice belong to the Crustacea, a group of invertebrates that includes crabs and shrimps. The only land-dwelling members of this group, they are important members of the litter fauna. They feed on leaves, and the process of digesting them and depositing them as faecal pellets speeds up the process of humus formation. *Porcellio scaber* is found in a range of woodland habitats, both damp and dry. The most common European woodlouse is *Oniscus asellus*, which prefers damp conditions, especially under the bark of decaying wood. *Philoscia muscorum* is more common at wood margins and in grassy clearings, and can be distinguished from the other two species by the broken outline of the rear part of its body.

ROUNDED SNAIL *Discus rotundatus*  GARLIC SNAIL *Oxychilus alliarius*

**SNAILS**
The leaf litter of an oakwood will contain a number of different species of small snail, especially where the soil is not too acid. The food of snails consists largely of decomposing leaves, which contain cellulose that the snails are able to digest. The Garlic Snail does smell distinctly of garlic, and both it and the Rounded Snail are common in oakwoods throughout Britain, but in mainland Europe the Garlic Snail has a western distribution.

*Philoscia muscorum* (SHOWN X 4)

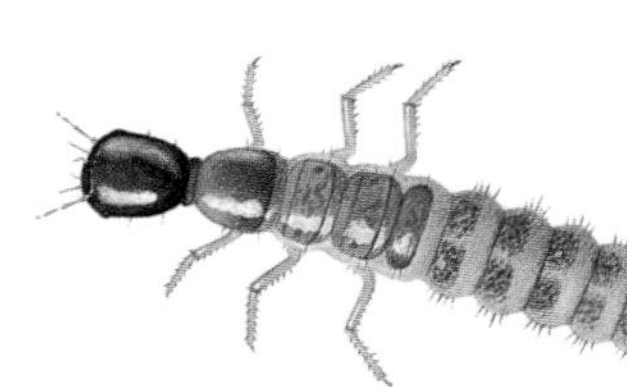

DEVIL'S COACH HORSE LARVA *Staphylinus olens* (SHOWN X 2½)

**DEVIL'S COACH HORSE**
***Staphylinus olens***
Rove beetles, or "cock-tail beetles", (Staphylinidae) resemble ground beetles *(see page 36)*, but have very short wing cases, which leave most of the abdomen exposed. Their habit of curling the tip of the abdomen over their backs when disturbed has earned them the name cock-tail beetle. The Devil's Coach Horse is the largest British rove beetle, and both the adult and the larva live in leaf litter and decaying vegetation and are active by night.

STAG BEETLE LARVA

**STAG BEETLE**
***Lucanus cervus***
The Stag Beetle is the largest European beetle. The larvae, which take about three years to reach maturity, are totally dependent on decaying wood and are found in the stumps and roots of oak and elm trees.

STAG BEETLE *Lucanus cervus*

**Spiders**

Spiders are an important part of the leaf litter fauna. The most numerous of all spiders are the money spiders. Most of these are tiny, and they catch their food in simple, sheet-like webs. *Macrargus rufus* is common in damp places under stones. The Lycosidae, or wolf spiders, are large, active predators. *Trochosa ruricola* is a species common under moss and in leaf litter. Rather than spin webs, wolf spiders hunt their prey in the upper layers of the leaf litter.

**False scorpions**

These fascinating little animals are not scorpions at all, but belong in a group of their own, the Pseudoscorpiones. They live in leaf litter, in moss, behind dead bark, or in birds' nests. All are small, the largest being no more than 4 mm (¼ in) long, with a pair of ridiculously large claws. They are carnivorous, preying on smaller members of the litter fauna. *Neobisium muscorum* is a widespread species in moss and woodland leaf litter.

**Wood Cricket**
***Nemobius sylvestris***

The Wood Cricket's song is one of the sounds of summer. Widespread in central and southern Europe, in Britain it is almost entirely confined to the old woodlands of the New Forest, where it feeds on leaf litter.

**Fungi**

The only organisms capable of breaking down wood, fungi play an essential role in the early stages of recycling fallen twigs and branches into humus. Sulphur-tuft Fungi and Many-zoned Polypores sprout on the stump shown here, and the Stag Beetle is crawling over a patch of the lichen *Cladonia coniocraea*.

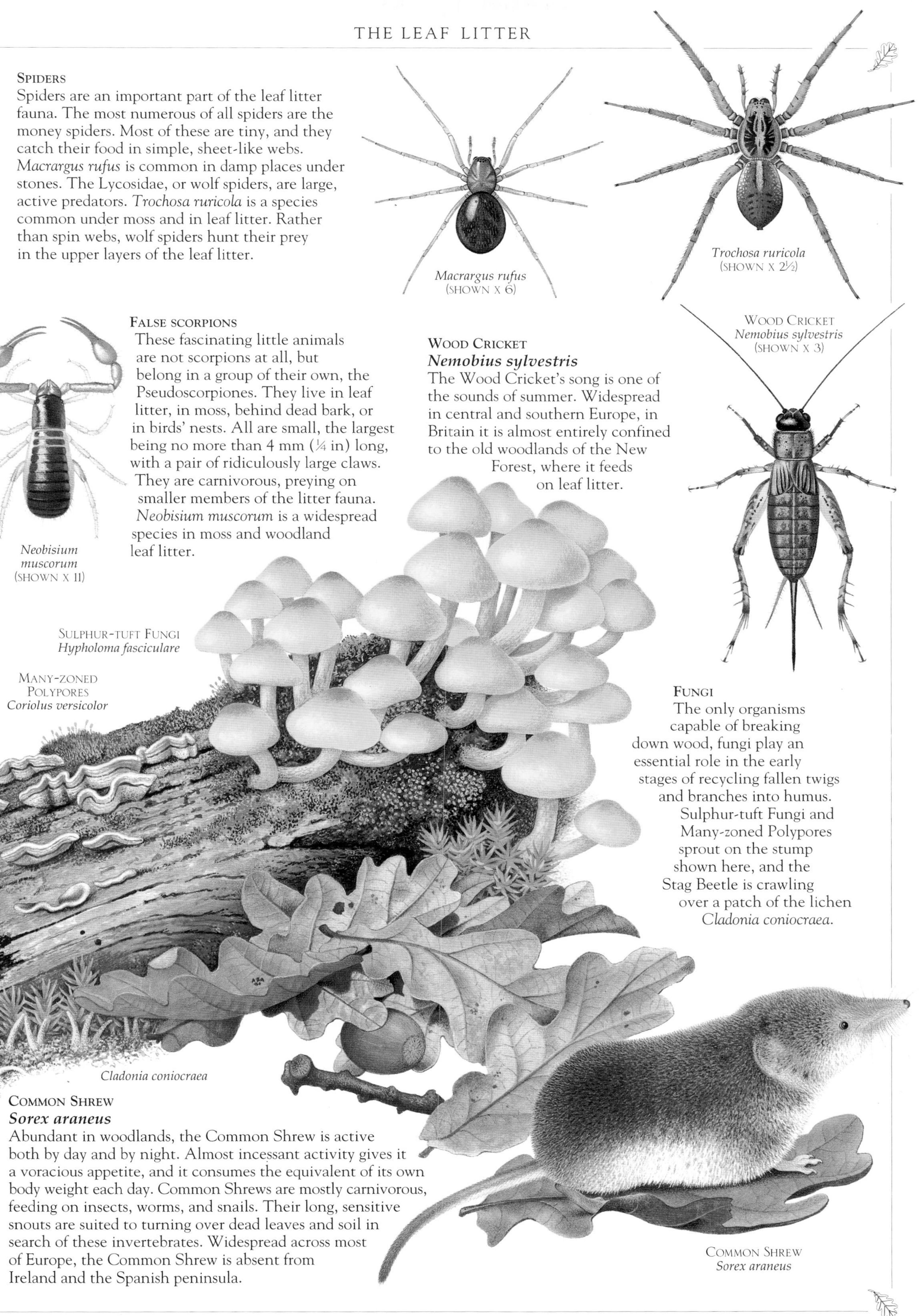


**Common Shrew**
***Sorex araneus***

Abundant in woodlands, the Common Shrew is active both by day and by night. Almost incessant activity gives it a voracious appetite, and it consumes the equivalent of its own body weight each day. Common Shrews are mostly carnivorous, feeding on insects, worms, and snails. Their long, sensitive snouts are suited to turning over dead leaves and soil in search of these invertebrates. Widespread across most of Europe, the Common Shrew is absent from Ireland and the Spanish peninsula.

# INDEX